Characterization, Testing, Measurement, and Metrology

Manufacturing Design and Technology Series

Series Editor:
J. Paulo Davim

This series will publish high-quality references and advanced textbooks in the broad area of manufacturing design and technology, with a special focus on sustainability in manufacturing. Books in the series should find a balance between academic research and industrial application. This series targets academics and practicing engineers working on topics in materials science, mechanical engineering, industrial engineering, systems engineering, and environmental engineering as related to manufacturing systems, as well as professions in manufacturing design.

Drills
Science and Technology of Advanced Operations
Viktor P. Astakhov

Technological Challenges and Management
Matching Human and Business Needs
Edited by Carolina Machado and J. Paulo Davim

Advanced Machining Processes
Innovative Modeling Techniques
Edited by Angelos P. Markopoulos and J. Paulo Davim

Management and Technological Challenges in the Digital Age
Edited by Pedro Novo Melo and Carolina Machado

Machining of Light Alloys
Aluminum, Titanium, and Magnesium
Edited by Diego Carou and J. Paulo Davim

Additive Manufacturing
Applications and Innovations
Edited by Rupinder Singh and J. Paulo Davim

For more information about this series, please visit: https://www.routledge.com/ Manufacturing-Design-and-Technology/book-series/CRCMANDESTEC

Characterization, Testing, Measurement, and Metrology

Edited by
Chander Prakash, Sunpreet Singh,
and J. Paulo Davim

CRC Press
Taylor & Francis Group
Boca Raton London New York

CRC Press is an imprint of the
Taylor & Francis Group, an **informa** business

First edition published 2021
by CRC Press
6000 Broken Sound Parkway NW, Suite 300, Boca Raton, FL 33487-2742

and by CRC Press
2 Park Square, Milton Park, Abingdon, Oxon, OX14 4RN

Library of Congress Cataloging-in-Publication Data
Names: Prakash, Chander, editor. | Singh, Sunpreet, editor. | Davim, J. Paulo, editor.
Title: Characterization, testing, measurement, and metrology / edited by Chander Prakash, Sunpreet Singh, J. Paulo Davim.
Description: First edition. | Boca Raton : CRC Press, 2020. | Series: Manufacturing design and technology | Includes bibliographical references and index.
Identifiers: LCCN 2020018334 (print) | LCCN 2020018335 (ebook) | ISBN 9780367275150 (hardback) | ISBN 9780429298073 (ebook)
Subjects: LCSH: Materials—Testing.
Classification: LCC TA410 .C457 2020 (print) | LCC TA410 (ebook) | DDC 620.1/10287—dc23
LC record available at https://lccn.loc.gov/2020018334
LC ebook record available at https://lccn.loc.gov/2020018335

ISBN: 978-0-367-27515-0 (hbk)
ISBN: 978-0-429-29807-3 (ebk)

Typeset in Times
by codeMantra

Contents

Preface

This book presents a comprehensive treatise of various operational principles, scientific tools, technical methodologies, and qualitative/quantitative characteristics involved in the scientific world. It is comprised four sections: (i) mechanical testing to reveal information about a material's mechanical properties under dynamic or static forces, (ii) metrological investigations involved to observe the geometrical and structural integrity of the engineering products using optical and analogue calipers, (iii) characterization of the engineering products using spectroscopic tools, and (iv) routine observations that play a vital role in quality assurance. And, measurements made within an academic institution, manufacturing facility, or research and development center must be able to be reproduced accurately anywhere in the world.

Therefore, the ultimate aim of this book is to understand the reliability aspects of the testing procedures, analyzed products, and quality issues of such procedures in order to understand their industrial implications. In addition, due to the overwhelming growth of the optics-based measurement protocols, characterization techniques such as metallography (light microscopy), X-ray diffraction, transmission and scanning electron microscopies, and the theoretical concept strength by using such protocols have been described.

We have also incorporated the research activities on the principles of advanced characterization and testing, including the importance of performance-based specifications in the manufacturing sector. Apart from the physical examination of the developed engineering products, measurement and testing of same via modeling and simulation also play a critical role to understand the design features and therefore have been incorporated in this edited book.

Editors

Chander Prakash is Associate Professor at the School of Mechanical Engineering, Lovely Professional University, Jalandhar, India. He received a Ph.D. in Mechanical Engineering from Panjab University, Chandigarh, India. His areas of research is biomaterials, rapid prototyping and 3D printing, advanced manufacturing, modeling, simulation, and optimization. He has more than 11 years of teaching experience and 6 years of research experience. He has contributed extensively to titanium- and magnesium-based implant literature with publications appearing in *Surface and Coating Technology, Materials and Manufacturing Processes, Journal of Materials Engineering and Performance, Journal of Mechanical Science and Technology, Nanoscience and Nanotechnology Letters*, and *Proceedings of the Institution of Mechanical Engineers, Part B: Journal of Engineering Manufacture*. He has authored 150 research papers and 30 book chapters. He is also an editor of 15 Books: He is also a guest editor of two journals: Special Issue on "Metrology in Materials and Advanced Manufacturing," *Measurement and Control* (SCI indexed) and Special Issue on "Nano-Composites and Smart Materials: Design, Processing, Manufacturing and Applications" of *Advanced Composites Letters*.

Sunpreet Singh is researcher in NUS Nanoscience & Nanotechnology Initiative (NUSNNI). He received a Ph.D. in Mechanical Engineering from Guru Nanak Dev Engineering College, Ludhiana, India. His area of research is additive manufacturing and application of 3D printing for development of new biomaterials for clinical applications. He has contributed extensively to the subject of additive manufacturing with publications appearing in *Journal of Manufacturing Processes, Composite Part: B, Rapid Prototyping Journal, Journal of Mechanical Science and Technology, Measurement, International Journal of Advance Manufacturing Technology*, and *Journal of Cleaner Production*. He has authored 10 book chapters and monographs. He is working in joint collaboration with Prof. Seeram Ramakrishna, NUS Nanoscience & Nanotechnology Initiative and Prof. Rupinder Singh, Manufacturing Research Lab, GNDEC, Ludhiana. He is also an editor of three books: *Current Trends in Bio-manufacturing*, Springer Series in Advanced Manufacturing, Springer International Publishing AG, Gewerbestrasse 11, 6330 Cham, Switzerland., December 2018; *3D Printing in Biomedical Engineering*, Book series Materials Horizons: From Nature to Nanomaterials, Springer International Publishing AG, Gewerbestrasse 11, 6330 Cham, Switzerland, August 2019; and *Biomaterials in Orthopaedics and Bone Regeneration - Design and Synthesis*, Book series: Materials Horizons: From Nature to Nanomaterials, Springer International Publishing AG, Gewerbestrasse 11, 6330 Cham, Switzerland, March 2019. He is also Guest Editor of three journals: Special Issue on "Functional Materials and Advanced Manufacturing," Facta Universitatis, Series: Mechanical Engineering (Scopus Index), Materials Science Forum (Scopus Index), and Special Issue on "Metrology in Materials and Advanced Manufacturing," *Measurement and Control* (SCI indexed).

J. Paulo Davim received a Ph.D. in Mechanical Engineering in 1997, a M.Sc. degree in Mechanical Engineering (materials and manufacturing processes) in 1991, a Mechanical Engineering degree (five years) in 1986 from the University of Porto (FEUP), the Aggregate title (Full Habilitation) from the University of Coimbra in 2005, and a D.Sc. from London Metropolitan University in 2013. He is Senior Chartered Engineer by the Portuguese Institution of Engineers with an MBA and Specialist title in Engineering and Industrial Management. He is also Eur Ing by FEANI-Brussels and Fellow (FIET) by IET-London. Currently, he is a professor at the Department of Mechanical Engineering of the University of Aveiro, Portugal. He has more than 30 years of teaching and research experience in manufacturing, materials, and mechanical and industrial engineering, with special emphasis in machining and tribology. He has also interest in management, engineering education, and higher education sustainability. He has guided many postdoc, Ph.D. and master's students as well as coordinated and participated in several financed research projects. He has received several scientific awards. He has worked as evaluator of projects for the European Research Council (ERC) and other international research agencies as well as examiner of Ph.D. candidates for many universities in different countries. He is the editor-in-chief of several international journals, a guest editor of journals, books series, and a scientific advisor for many international journals and conferences. Presently, he is an editorial board member of 30 international journals and acts as reviewer for more than 100 prestigious Web of Science journals. He has also published as editor (and co-editor) of more than 100 books and as author (and co-author) of more than 10 books, 80 book chapters, and 400 articles in journals and conferences (more than 250 articles in journals indexed in Web of Science core collection/h-index 49+/7000+ citations, SCOPUS/h-index 56+/10000+ citations, Google Scholar/h-index 70+/16000+).

Contributors

Atul Babbar
Mechanical Engineering Department
Thapar Institute of Engineering and
    Technology
Patiala, India

Dr. Harish Kumar Banga
Department of Production & Industrial
    Engineering
Punjab Engineering College
Chandigarh, India

Monoj Bardalai
Department of Mechanical Engineering
Tezpur University
Assam, India

Dr. R.M. Belokar
Department of Production & Industrial
    Engineering
Punjab Engineering College
Chandigarh, India

Vibhanshu Chhettri
Department of Mechanical Engineering
DIT University
Dehradun, India

V.K. Dwivedi
GLA University
Mathura, India

Bhaskarjyoti Gogoi
Department of Mechanical Engineering
Tezpur University
Assam, India

Rajeev Goswami
Department of Mechanical Engineering
Tezpur University
Assam, India

Ankit Gupta
Shiv Nadar University
Dadri, India

Dheeraj Gupta
Mechanical Engineering Department
Thapar Institute of Engineering and
    Technology
Patiala, India

Nitin Kumar Gupta
Department of Mechanical Engineering
DIT University
Dehradun, India

Vivek Jain
Mechanical Engineering Department
Thapar Institute of Engineering and
    Technology
Patiala, India

Dr. Parveen Kalra
Department of Production & Industrial
    Engineering
Punjab Engineering College
Chandigarh, India

Dr. Rajesh Kumar
Department of Mechanical
    Engineering
UIET, Panjab University
Chandigarh, India

Yogesh Kumar
Shiv Nadar University
Dadri, India

Siba Sankar Mahapatra
National Institute of Technology
    Rourkela
Rourkela, India

Manoj Mittal
Department of Mechanical Engineering
IKG Punjab Technical University
Jalandhar, India

Praveen Kumar Nayak
National Institute of Technology
    Rourkela
Rourkela, India

Chander Prakash
School of Mechanical Engineering
Lovely Professional University
Phagwara, India

Anshuman Kumar Sahu
National Institute of Technology
    Rourkela
Rourkela, India

Shahabuddin
GLA University
Mathura, India

Akash Sharma
GLA University
Mathura, India

Ankit Sharma
Chitkara College of
    engineering
Chitkara University
Patiala, India

Dheer Singh
Shiv Nadar University
Dadri, India

Pawan Singh
Department of Mechanical
    Engineering
DIT University
Dehradun, India

Sunpreet Singh
School of Mechanical
    Engineering
Lovely Professional University
Phagwara, India

Md. Adam Yamin
Department of Mechanical
    Engineering
Tezpur University
Assam, India

1 Wear Measuring Devices for Biomaterials

Manoj Mittal
IKG Punjab Technical University Jalandhar

CONTENTS

1.1   WEAR OF BIOMATERIALS: HOW IT IS DIFFERENT FROM OTHER MATERIALS

The failure of body implant may be due to wear of one of the joining parts in the body environment. The wear in the implant and surrounding bone may be abrasive, adhesive or fatigue. Some wear testing devices are used to simulate the wear in body environment. Using these wear testing machines, a highly accelerated wear environment (as compared to actual body environment) is created for a shorter time span than actual service life of implant. By doing so, the best suited implant material may be recommended by testing the same in laboratories (in vitro). Wear of biomaterials is measured using specially designed simulators. These simulators simulate actual body environment, and original body implants are tested in these machines for lakhs of cycles in a shorter time span than the actual life span of implant. This is done by highly increasing the number of cycles per unit time.

Knee simulator is one of the most complex wear testing devices, which tests actual knee prostheses, and is very costly. Some commercial equipment may cost as expensive as $200,000 or even more. Wang et al. (1999) and Blunn et al. (1991) have reported a comparatively less expensive substitute for knee simulators to examine materials used for knee replacement. Other simpler and inexpensive models for wear testing are pin-on-flat (Van Citters, 2004), pin-on-disc, flat-on-disc and cyclic sliding wear testing machines. Morks et al. (2007) utilized SUGA abrasion tester, which follows NUS-ISO-3 standards (Japan) for testing wear on bio-inserts; a similar wear testing machine is built to calculate the wear resistance of coated specimens by Mittal (2012).

The release of debris due to wear and subsequent tissue inflammatory response has emerged as a central problem, restraining the long-term clinical outcome of total hip replacements (Harris, 1995; MaGee et al., 1997; Wroblewski, 1997; Raimondi, 2001). The key wear mechanisms observed on retrieved knee prosthesis include delamination caused by surface damage of polyethylene, surface pitting, third body wear and adhesive wear (Blunn et al., 2009; Hood et al., 1983; Landy and Walker, 1988; Collier et al., 1991). Low conformity designs were found to be the cause of delamination (Engh et al., 1992), whereas in a relative conforming design, surface damage was found to be associated with entrapped acrylic particles (Hood et al., 1983). The damage may be because of the different kinematic conditions occurring at bearing surface (Blunn et al., 2009). Excessive sliding led to delamination wear, whereas rolling or cyclic loading at the same contact point resulted in minimal wear (Blunn et al., 1991).

Three body wear and production of polyethylene and metallic debris generally occur mainly at articulating joint; however, a little may occur at femoral stem. Mechanical stresses generated by patient on hip implant are supposed to be the foundation of the third body wear. There is a mixed response on the effectiveness of hydroxyapatite (HA) coatings in preventing third body wear. Several clinical studies have revealed that HA coatings had no adverse effects; however, other clinical studies have discovered excessive wear at the polyethylene surface due to the accumulation of calcium phosphate and metal particles due to third body wear (Sun et al., 2001).

Shearing micro-movements may take place at implant and bone interface due to a large difference in elastic modulus of two materials in contact. Insufficient initial fixation (problem in prosthesis design) or movement of limb (which sustains many stresses) in some course of time can also cause micro-movement (Fu, 1999; Walker et al., 1987; Riues et al., 1995). The oscillatory micro-movements at the contact induce fretting wear, fretting corrosion and sometimes fretting cracks, causing early failure of joint prosthesis (Hoppner and Chandrasekaran, 1994; Lambardi et al., 1989). In an investigation by Gross and Babovic (2002), abrasion resistance of coatings was ascertained with pin-on-disc arrangement under unlubricated conditions. A bone analogue made of wood, with φ6.3, was used as pin to simulate cortical bone in terms of hardness and elastic modulus. The result of the investigation showed a weight loss of coating and decrease in surface roughness, and main change in the surface characteristics occurred in first minute of testing (Gross and Babovic, 2002).

Coathup et al. (2005) inserted six different types of hip replacements (36 in total) into the right hip of skeletally mature female mule sheep and revealed that HA-coated implants were more effective than other uncemented and cemented implants in resisting progressive osteolysis along the acetabular cup–bone interface and also concluded the importance of the in-growth surface present on implant. HA-coated porous acetabular implants showed significant results in terms of bone contact and in-growth in the presence of wear debris and in the prevention of interfacial wear particle migration.

Kalin et al. (2003) studied the wear of the hydroxyapatite pins against glass-infiltrated alumina submerged in a static bath of distilled water at room temperature and reported that wear of hydroxyapatite pin against glass-infiltrated alumina occurred primarily by fracture and deformation. The hydroxyapatite wear particulates are mixed with wear products from glass infiltrate in alumina and water to form an intermediate surface layer. Because of this adhered debris layer, steady-state wear is more appropriately described as three-body wear as opposed to two-body wear. Pin-on-disc experiments with HA pins and glass-infiltrated alumina (in-Ceram alumina) conducted by Kalin et al. (2002) showed that the wear volume of HA increased as surface roughness of glass-infiltrated alumina and load was increased, while for a given surface roughness value, the wear factor remained independent of load. Furthermore, polished glass-infiltrated surface showed no evidence for material transfer at low load, whereas mechanical wear with removal of glass infiltrate was observed at higher loads.

Morks et al. (2007) investigated the role of arc current in plasma spray technique on abrasion behavior of coatings and reported that with an increase in arc current, the abrasion resistance of HA coating increases mainly due to the increase in hardness of coating. The resistance to abrasion wear was found to be dependent on coating thickness because the abrasion wear resistance increased as the thickness of HA coating becomes less than $30\,\mu m$ due to the increase in hardness of thin HA coatings. Morks and Kobayashi (2006) studied the dependence of gas flow rate on plasma-sprayed HA coatings and reported that HA coatings sprayed at high flow rates exhibit higher abrasive wear resistance compared to those sprayed at low gas flow rate due to higher cohesion bonding among the splats and low porosity.

The wear resistance of HA can be enhanced by reinforcing the secondary phase to HA to produce composite coatings. Several researchers have used various reinforcement materials such as silica (SiO_2), titania (TiO_2) (Morks, 2008), alumina (Al_2O_3), zirconia (ZrO_2), carbon nanotubes (CNTs), diamond-like carbon, P_2O_5–CaO glass, yttria-stabilized zirconia (YSZ) (Balani, 2007), Ni_3Al, and titanium and its alloys. A composite powder of HA with 4 wt.% multi-walled CNTs was deposited on Ti-6Al-4V. Both HA and HA-CNT composite coatings showed better wear resistance than Ti-6Al-4V substrate, whereas HA-CNT composite coatings result in reduced weight and volume loss in comparison with HA coatings and Ti-6Al-4V substrate. Low weight loss of HA-CNT coating during wear was due to the under-propping and self-lubricating nature of CNTs and the pinning of wear debris assisted by CNT bridging and stretching (Balani et al., 2007). The enhancement in wear resistance (Tercero et al., 2009) was observed by reinforcing CNT to HA; furthermore, the resistance to wear was increased by increasing the content of CNT from 0% to 20% and this behavior (Chen, 2007) might be attributed due to the increased hardness, strength, and fracture toughness of composite coatings (Lahiri, 2011) compared to pure HA coating (Chen et al., 2007). Alumina offers a very high wear resistance at articulating surface in orthopedic applications due to its high hardness, low coefficient of friction and excellent resistance to corrosion (Cordingley et al., 2003). Wang et al. (2005) had examined the wear properties with respect to partially stabilized ZrO_2 reinforcement to HA against UHMWP in human plasma lubrication and reported improvement in resistance to wear might be due to the addition of reinforced particles. In HVOF-sprayed HA/TiO_2 composite, mutual reaction could be the cause of chemical bonding between HA and titanium splats. The chemical bonding was found to be beneficial for the prevention of release of titanium particles as wear debris which can lead to prosthesis rejection or infection (Li et al., 2002).

1.2 MACHINES AND EQUIPMENT FOR BIOMATERIALS: UNDER DRY CONDITIONS AND IN SIMULATED BODY ENVIRONMENT JOINT SIMULATORS

Joint simulators are standard testing equipment used to measure tribological behavior of actual joint used in repair and replacements of broken and damaged bone and tissues in an artificially created environment. They mimic the orientation and magnitude of the load of actual joint. Ideal simulators are sealed and temperature controlled, and produce a similar type of wear with a similar rate of wear debris that is of comparable morphology as found in actual clinical cases. After completing millions of cycles in a simulator (which take a considerable time), specimens are weighed and the difference in weight loss is measured by comparing with the initial weight of the specimen. A simulator, in general, does not provide quantitative data directly. The wear is also inspected with a scanning electron microscope. These simulators only account for physical tribological factors and do not take into consideration the biological factors that would affect the tribology in a body. Biological factors may be taken care of by introducing simulating/artificial body environment in the simulator.

The choice of lubricant to be used in joint simulator is another important factor to be considered. Joint simulator requires a relatively large quantity of lubricant fluid. The human joint fluid is not available due to its limited supply, as healthy knees do not contain much fluid and due to inflammation, the knees have enough synovial fluid for suitable extraction. The research studies thus relegated to use synthetic lubricants; American Society for Testing and Materials (ASTM) simply recommends that these lubricants be volume, concentration and temperature controlled (Hirakawa et al., 2004). ASTM recommends a bovine serum lubricant supplemented with 0.2%–0.3% sodium azide and 20 ml ethylenediaminetetraacetic acid (EDTA). The sodium azide limits bacterial growth in the fluid and has a role in inhibiting protein adsorption on the surface of implant, whereas EDTA discourages calcium phosphate precipitation. Overall, a joint simulator provides a reasonably good idea regarding wear performance for artificial joints. The time taken for analysis, variability of lubricants, and, foremost, the cost limit their use. Unfortunately, the cost of commercial models of knee simulator exceeds $200,000 (₹10 million).

The most complex wear testing devices are knee simulators, which test actual knee prosthesis. Most of the knee simulators provide physiologic loading and can reproduce at least four out of six degrees of freedom: flexion/extension (F/E), anterior/posterior (A/P), tibial rotation and abduction/adduction. Some of the joint simulators are discussed in detail hereunder.

1.2.1 SIX-STATION ROLLING SLIDING TRIBO-TESTER

The six-station rolling sliding tribo-tester is a quite effective machine (Kennedy et al., 2006) for testing the contact fatigue behavior of different orthopedic poly-mers (Van Citters et al., 2004) and is equally useful in testing their wear behavior due to clinically relevant stress and motion environments. The tribo-tester articu-lates cobalt–chromium cylinders against UHMWPE pucks (Kennedy, 2006). The Hertzian line contact theory is utilized to determine contact stresses resulting from specimen geometry. Moreover, the non-confirming contact creates surface subsur-face stresses that are similar to those found in contemporary total knee replacement bearings (Kennedy et al., 2000 and Van Citters et al., 2004). Alignment of the articu-lating cylinders is performed by placing pressure-sensitive films at contact between a Co-Cr cylinder and an UHMWPE puck. The macrograph of rolling sliding tribo-tester is represented in Figure 1.1.

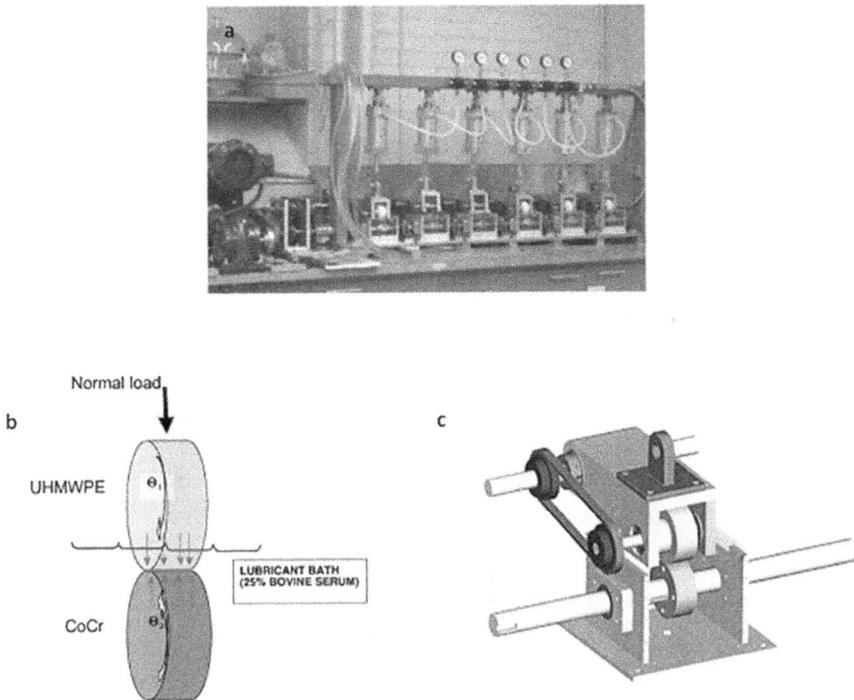

FIGURE 1.1   Multi-station rolling sliding tribo-tester: (a) macrograph of tribo-tester, (b) con-tact conditions in one of the stations (Kennedy et al., 2007) and (c) CAD model of one of the six stations. UHMWPE specimen is attached to upper shaft, whereas metallic specimen is with lower shaft. The contact region is contained in a sealed fluid chamber (Van Citters, 2004).

The intensity of color on otherwise translucent film reflects contact pressure at that location. Shims are generally inserted in the bearing mounts to make the Co-Cr cylinder and UHMWPE puck parallel, and by doing so, uniform contact pressure profile is created. After the insertion of shims, all six stations should have the same distribution of pressure across the contact width. Each station has a fluid bath, which provides lubrication and cooling at contact points. The contact can be viewed through the acrylic windows incorporated with the system. The design of the testing equipment ensures that all contact axes are normal to direction of sliding. Each pair can be loaded independently as per the requirement of test as constant load pneumatic system is incorporated with each pair of cylinder and puck. A crank and rocker arm powered by a fixed-speed motor is used for articulation. Backlash is prevented by adjustable gears as well as adjustable chains and sprockets. The degree of sliding can be adjusted by combinations of sprockets in the drive train. The two test specimens can oscillate and rotate with different amplitudes and speeds.

The relative oscillation can be altered for each pair of specimens independently as per the requirement of test. Linear wear can be determined by using profilometer. The measurements should be taken at the earliest after the test has been completed to avoid any viscoelastic relaxation. A six-station knee simulator is represented in Figure 1.2.

1.2.2 HOW-MEDICA BIAXIAL LINE CONTACT WEAR MACHINE

A wear machine with two parallel axes for evaluation of wear of implant bearing materials for total knee joint replacement was developed by Wang et al. (1999). The design used by the researchers is based on simulating both geometry and motion of knee joint.

FIGURE 1.2   Three stations of AMTI-Boston six-station knee simulator, Model KS2–6–1000 (AMTI, 200).

A static load of 1150 N is applied for simplicity as well as to reduce cost. A rotating Co-Cr ring (ϕ71.9$\times$25 mm) is loaded against UHMWPE block (1.6 cm$\times$1.6 cm$\times$1.25 cm), which oscillates about load axis to simulate tibial rotation. The degree of flexion-extension is 0°–60°, and that of tibial rotation is 0°–30°. Bovine serum is used to lubricate and cool the contact points of the specimens.

The limitation of this machine is the absence of anterior-posterior sliding and physiological correct loading. The static loading may decrease the wear rate. The wear testing equipment described here, which is very low in cost as compared to commercial six-station knee simulator manufactured by MTC (Eden Prairie, MN, USA) and Instron-Stanmore (Canton, MA, USA), shows parallel results of wear rates and wear mechanism in terms of morphology of wear surface and wear products. The macrograph of two of the twelve stations of How-Medica biaxial line contact wear machine (Wang et al., 1999) is represented in Figure 1.3.

1.2.3 THREE-AXIS KNEE WEAR SIMULATOR

The three-axis knee simulator with ball-on-surface contact was designed and built by Saikko et al. (2001) for basic wear and friction tests of materials used in knee prosthesis. The motion of machine consists of flexion-extension (F/E), anterior-posterior (A/P) translation and inward-outward (I/O) rotation. The F/E is applied by ball and the A/P and I/O with the disc, as represented in Figure 1.4a.

A crank mechanism is utilized to implement motion, and variation with time is nearly sinusoidal with a cycle time of 0.93 s. A large Co-Cr ball represents femoral component, and a flat polyethylene disc, located horizontally beneath the ball, represents tibial component, as represented in Figure 1.4b. The frame, motor and loading system of an old five-station, uniaxial hip joint simulator was utilized to complete one station of the three-axis knee wear simulator (and Saikko et al., 2001).

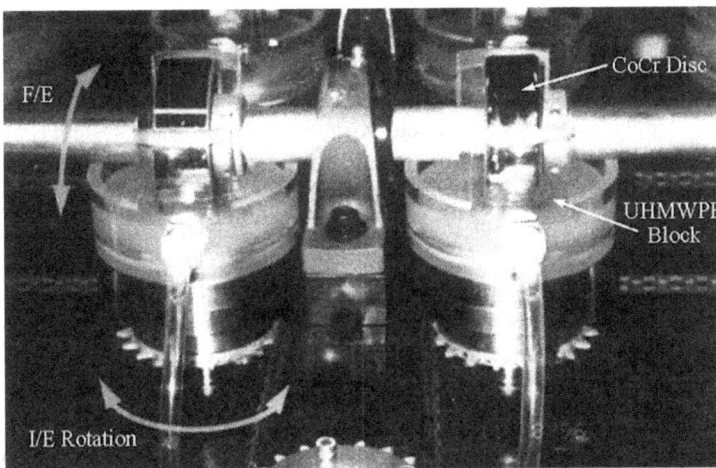

FIGURE 1.3 Two stations of twelve-station How-Medica biaxial line contact wear machine (Wang et al. 1998).

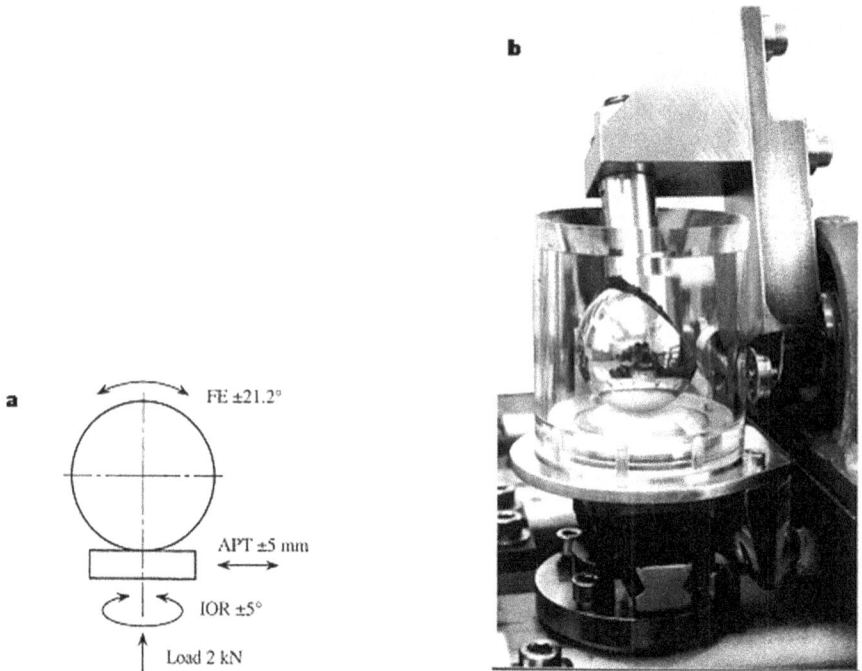

FIGURE 1.4 Three-axis knee simulator with ball-on-flat contact: (a) principle and (b) close-up.

The 54 mm diameter Co-Cr ball with 42.4° F/E motion was used to revolve on the disc, and a vertical upward static load of 2 kN was applied on the lower side of the disc, as shown in Figure 1.4. The detailed design of machine is reported by Saikko et al. (2001).

1.2.4 FEIN FOCUS X-RAY MICROSCOPE AND FRETTING WEAR APPARATUS

The equipment consists of a Fein Focus X-ray image processing system (Fein Focus Roentgen-System, Germany) with a beam spot size 4–20 μm in diameter (Fu et al., 1998). Micro-focused X-rays directly penetrate contact surfaces and form an evident image on the fluorescent screen placed below the specimen.

A high-speed digital camera captures the images developed on fluorescent screen and converted them to visual images, which can be displayed on a monitor. A definite time interval is selected to capture the images, and image processing functions are used to treat and enhance the images to reveal the detailed phenomenon of fretting wear. The line diagram of the equipment is represented in Figure 1.5.

This equipment is an improved version of simple ball-on-flat fretting wear testing machine. The upper specimen is a ball made from hardened stainless steel AISI 410 with a diameter of 25 mm, as shown in Figure 1.5.

The material to be investigated for wear studies is made in form of a flat and mounted on a table which reciprocates at a specified frequency. Electromagnetic

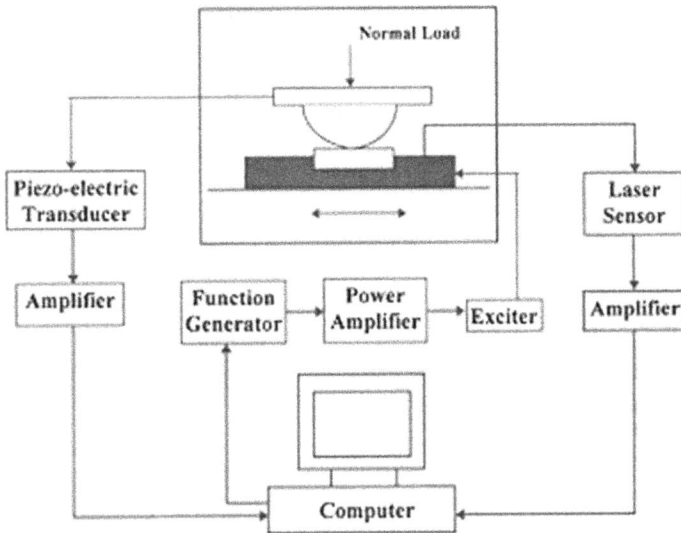

FIGURE 1.5 Schematic illustration of fretting wear tribometer (Fu et al., 1998).

exciters are used to oscillate the table at a given frequency. A function generator controls the displacement via a power amplifier (Fu, 1999). A piezoelectric force transducer is used to measure tangential frictional force, and the amplitude of oscillatory movement is measured with a laser sensor with a minimum resolution of $2 \mu m$ (Fu et al., 1998).

1.2.5 WEAR TESTING EQUIPMENT

All the joint simulators which simulate motions and loading of the actual human knee joint are very expensive for procurement and very difficult to handle and manage. Many investigators have used some simple machines to estimate the wear loss of the coated metallic as well as bare metallic specimens. Khan et al. (1996), Balani et al. (2007), Kalin et al. (2003) and Kalin et al. (2002) investigated wear resistance of specimens on pin-on-disc wear testing equipment using phosphate-buffered saline, simulated body fluid, carboxymethyl cellulose in distilled water, and distilled water, which is used to lubricate the contact, respectively, whereas Gross and Babovic (2002), Sidhu et al. (2007) and Bolelli et al. (2006) conducted wear experiments under dry conditions on pin-on-disc wear tester. Lima et al. (2006), Morks and Kobayashi (2006), Morks et al. (2007) and Morks and Akimoto (2008) used flat-on-disc without any lubricant to conduct wear experiments of coated specimens.

The pin-on-disc apparatus allows a pin to articulate on a smooth disc, which spins in a rotary manner beneath the pin including relative sliding motion between pin and disc. Normal load is applied, and frictional force in the transverse direction and tangent to circular track is obtained. This device provides both wear and frictional data. The pin can be weighted for loss of mass over a set period of revolutions or distance slid. The contact of pin and disc can be lubricated and cooled using some lubricant as discussed earlier. The reverse combination of pin and disc can also be used in which pin is made of harder material such as zirconia and disc is the specimen for which the wear resistance is to be measured. In this arrangement, disc can be weighed for any loss of mass. A schematic illustration of pin-on-disc apparatus is represented in Figure 1.6.

Karanjai et al. (2008) studied the fretting wear on Ti-Ca-P bio-composite against bearing steel over 10,000 cycles at a frequency of 10 Hz and stroke of 80 μm with 2–10 N load under dry conditions and in Simulated Body Fluid (SBF), and reported that under dry conditions, the steady-state coefficient of friction increases with load, whereas in SBF, it is independent of load and have a very low value when compared to dry conditions. The fretting wear rates were found to be less by two orders of magnitudes in SBF than those under dry conditions at low loads. Under dry conditions, the wear rate decreased at high loads due to the formation of a tribo-layer, whereas it increased with load in SBF medium.

The flat-on-disc allows a flat surface to slide on the periphery of a disc, which revolves beneath the flat, as represented in Figure 1.7. The rpm of the disc can be changed by means of gear attachment. A servo motor is incorporated to slide the test sample as well as to rotate the disc about its axis. A vertical load is applied. The weight loss of flat specimen is measured to calculate the wear if the disc is an abrasive wheel, whereas the weight loss of flat and disc is measured if both are coated.

As shown in Figure 1.7 in SUGA abrasion testing equipment, the 1 cm wide wheel is covered with 400 grit SiC emery paper, which moves at 25 mm/min. The flat slides 1 cm on the disc with a normal load of 10 N. The flat specimen and the disc have a line contact of 1 cm (Morks et al., 2007, 2008; Morks and Akimoto, 2008). The SUGA wear tester follows NUS-ISO-3 standard (Japan).

Inspired by the design which offers better wear measurement due to line contact, combination of sliding and rotational motion, which reproduces two degrees

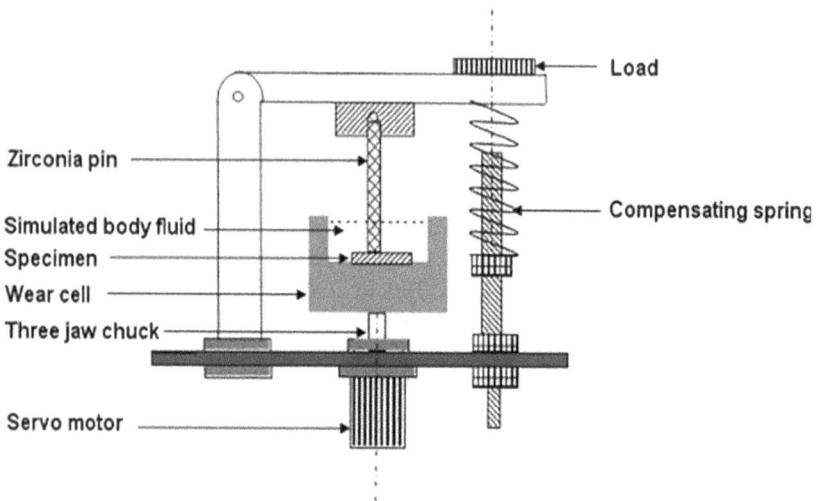

FIGURE 1.6 Schematic illustration of pin-on-disc wear machine with lubrication facility (Balani et al., 2007).

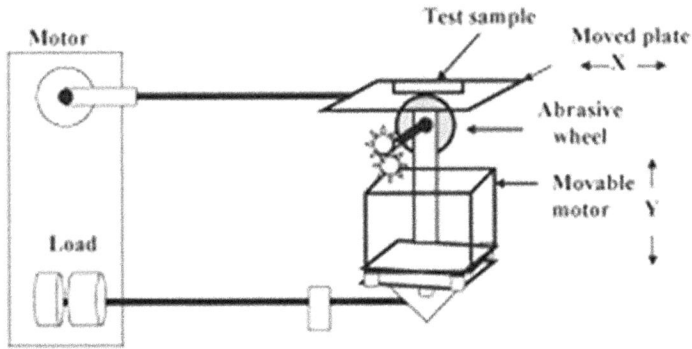

FIGURE 1.7   Schematic illustration of flat-on-disc wear tester (SUGA wear tester) (Morks et al., 2007).

of motion, i.e., anterior-posterior and flexion-extension (flat sliding specimen and rotating disc) out of six degrees of motion of knee joint, similar equipment was built. Flannery et al. (2008) simulated the same two degrees of motion in three-station wear simulators to measure wear of tibial inserts.

1.3   CONCLUSIONS

Various types of wear testing machines are discussed in detail, and it was found that for actual body implants, high-end multi-axes simulators with many degrees of freedoms are very expensive and can only be utilized for testing of real-time body implants in manufacturing industries. Other similar low-cost machines are suitable for laboratory work.

1.4   FUTURE SCOPE

Different types of wear testing equipment and machines are extensively used by researchers as well as in medical industry. In these machines and equipment, small-scale replica of actual implant with the same or similar materials is tested in an aggressive environment. Real implants need to be tested, and machines are being developed which can measure life span of real implants in an artificial real-like environment and different loading conditions of various limbs of human body. New and better reinforcements are added to increase the service life of implant.

REFERENCES

AMTI.biz.
Balani, K., Anderson, R., Laha, T., Andara, M., Tercero, J., Crumpler, E. and Agarwal, A., (2007), "Plasma-sprayed carbon nanotube reinforced hydroxyapatite coatings and their interaction with human osteoblasts in vitro", *Biomaterials*, Vol. 28, pp. 618–624.
Balani, K., Chen, Y., Harimkar, S. P., Dahotre, N. B. and Agarwal, A., (2007) "Tribological behavior of plasma sprayed carbon nanotube-reinforced hydroxyapatite coating in physiological solution", *Acta Biomater.*, Vol. 3, pp. 944–951.

Blunn, G. W., Joshi, A. B., Lilley, P. A. and Engelbrecht, E., (2009), "Polyethylene wear in unicondylar knee prostheses", *Acta Orthop. Scand.*, Vol. 63, pp. 247–255.

Blunn, G. W., Walker, P. S., Joshi, A. and Hardinge, K., (1991), "The dominance of cyclic sliding in producing wear in total knee replacements", *Clin. Orthop.*, Vol. 273, pp. 253–260.

Bolelli, G., Cannillo, V., Lusvarghi, L. and Manfredine, T., (2006), "Wear behavior of thermally sprayed ceramic oxide coatings", *Wear*, Vol. 261, pp. 1298–1315.

Chen, Y., Zhang, T. H., Gan, C. H. and Yu, G., (2007), "Wear studies of hydroxyapatite composite coating reinforced by carbon nanotubes", *Carbon*, Vol. 45, pp. 998–1004.

Coathup, M. J., Blackburn, J., Goodship, A. E., Cunningham, J. L., Smith, T. and Blunn, G. W., (2005) "Role of hydroxyapatite coating in resisting wear particle migration and osteolysis around acetabular components", *Biomaterials*, Vol. 26, pp. 4161–4169.

Collier, J. P., Mayor, M. B., McNamara, J. L., Suprenant, V. A. and Jensen, R. E., (1991), "Analysis of failure of 122 polyethylene inserts from uncemented tibial knee components", *Clin. Orthop.*, Vol. 273, pp. 232–242.

Cordingley, R., Kohan, L., Ben-Nissan, B. and Pezzotti, G., (2003), "Alumina as an orthopedic biomaterial – characteristics, properties, performance and applications", *J. Aust. Ceram. Soc.*, Vol. 39 (1), pp. 20–28.

Dorozhkin, S. V., (2016), "Biocomposites and hybrid biomaterials based on CaPO4", In *Calcium Orthophosphate-Based Bioceramics and Biocomposites.* doi:10.1002/9783527699315 Wiley-VCH Verlag GmbH & Co. KGaA

Engh, G. A., Dwyer, K. A. and Hanes, C. K., (1992), "Polyethylene wear of metal backed tibial components in total and unicompartmental knee prostheses", *J. Bone Joint Surg. (Br)*, Vol. 74 (1), pp. 9–17.

Flannery, M., McGloughlin, T., Jones, E. and Birkinshaw, C., (2008), "Analysis of wear and friction of total knee replacements part I. Wear assessment on a three station wear simulator", *Wear*, Vol. 265, pp. 999–1008.

Fu, Y. Q., Batchelor, A. W. and Loh, N. L., (1998), "Revealing the hidden world of fretting wear processes of surface coatings by x-ray mapping", *Surf. Coat. Technol.*, Vol. 107, pp. 133–141.

Fu, Y., Batchelor, A. W. and Khor, K. A., (1999), "Fretting wear behavior of thermal sprayed hydroxyapatite coating lubricated with bovine albumin", *Wear*, Vol. 230, pp. 98–102.

Gross, K. A. and Babovic, M., (2002), "Influence of abrasion on the surface characteristics of thermally sprayed hydroxyapatite coatings", *Biomaterials*, Vol. 23, pp. 4731–4737.

Harris, W. H., (1995), "The problem in osteolysis", *Clin. Orthop.*, Vol. 311, pp. 46–53.

Hirakawa, K., Jacobs, J. J., Urban, R. and Saito, T., (2004), "Mechanism of failure of total knee replacements – lessons learned from retrieval studies", *Clin. Ortho.*, Vol. 420, pp. 10–17.

Hood, R. W., Wright, T. W. and Burstein, A. H., (1983), "Retrieval analysis of total knee prostheses a method and its application to 48 total condylar prostheses", *J. Biomed. Mater. Res.*, Vol. 17, pp. 829–842.

Hoppner, D. W. and Chandrasekaran, V., (1994), "Fretting in orthopedic implants: a review", *Wear*, Vol. 173, pp. 189–197.

Kalin, M., Hockey, B. and Jahanmir, S., (2003), "Wear of hydroxyapatite sliding against glass-infiltrated alumina", *J. Mater. Res.*, Vol. 18 (1), pp. 27–36.

Kalin, M., Jahanmir, S. and Ives, L. K., (2002), "Effect of counterface roughness on abrasive wear of hydroxyapatite", *Wear*, Vol. 252, pp. 679–685.

Karanjai, M., Manoj Kumar, B. V., Sundaresan, R., Basu, B., Rama Mohan, T. R. and Kashyap, B. P., (2008), "Fretting wear study on Ti-Ca-P biocomposite in dry and simulated body fluid", *Mater. Sci. Eng. A*, Vol. 475, pp. 299–307.

Kennedy, F. E., Currier, J. H., Plumet, S., Duda, J. L., Gestwick, D. P., Collier, J. P., Currier, B. H. and Dubourg, M. C., (2000), "Contact fatigue failure of ultra-high molecular weight polyethylene bearing components of knee prostheses", *ASME J. Tribol.*, Vol. 122, pp. 332–339.

Kennedy, F. E., Van Citters, D. W., Wongseedakaew, K. and Mongkolwongrojn, M., (2006), "Lubrication and wear of artificial knee joint materials in a rolling/sliding tribo-tester", *Part B: Magnetic Storage Tribology; Manufacturing/Metalworking Tribology; Nanotribology; Engineered Surfaces; Biotribology; Emerging Technologies; Special Symposia on Contact Mechanics; Special Symposium on Nanotribology.*

Kennedy, F. E., Van Citters, D. W., Wongseedakaew, K. and Mongkolwongrojn, M., (2007), "Lubrication and wear of artificial knee joint materials in a rolling/sliding tribotester", *Journal of Tribology*, Vol. 129, pp. 326–335.

Khan, M. A., Williams, R. L. and Williams, D. F., (1996), "In-vitro corrosion and wear of titanium alloys in the biological environment", *Biomaterials*, Vol. 17, pp. 2117–2126.

Lahiri, D., Singh, V., Benaduce, A. P., Seal, S., Kos, L. and Agarwal, A., (2011), "Boron nitride nanotube reinforced hydroxyapatite composite: mechanical and tribological performance and in-vitro biocompatibility to osteoblasts", *J. Mech. Behav. Biomed.*, Vol. 4, pp. 44–56.

Lambardi, A. L., Mallory, T. H., Vaughn, B. K. and Drouillard, P., (1989), "Aseptic loosening in total hip arthroplasty secondary to osteolysis induced by wear debris from titanium alloy modular heads", *J. Bone Joint Surg. Am.*, Vol. 71, pp. 1337–1342.

Landy, M. M. and Walker, P. S., (1988), "Wear of ultra high molecular weight polyethylene components of 90 retrieved knee prostheses", *J. Arthroplasty (Suppl.)*, Vol. 3, pp. S73–S85.

Li, H., Khor, K. A. and Cheang, P., (2002), "Titanium dioxide reinforced hydroxyapatite coatings deposited by high velocity oxy-fuel (HVOF) spray", *Biomaterials*, Vol. 23, pp. 85–91.

Li, H., Khor, K. A. and Cheang, P., (2003) "Impact formation and microstructure characterization of thermal sprayed hydroxyapatite/titania composite coatings", *Biomaterials*, Vol. 24, pp. 949–957.

Lima, R. S. and Marple, B. R., (2006), "From ASP to HVOF spraying of conventional and nanostructured titania feedstock powders: a study on the enhancement of the mechanical properties", *Surf. Coat. Technol.*, Vol. 200, pp. 3428–3437.

MaGee, M. A., Howie, D. W., Neale, S. D., Haynes, D. R. and Pearcy, M. J., (1997), "The role of polyethylene wear in joint replacement failure", *Proc. Instn. Mech. Engrs., Part H, J. Eng. Med.*, Vol. 211 (H1), pp. 65–72.

Mittal, M., (2012), "Wear and corrosion behaviour of plasma sprayed HA reinforced coatings", PhD. Dissertation, Indian Institute of Technology Roorkee, Roorkee, India.

Morks, M. F. and Akimoto, K., (2008), "The role of nozzle diameter on the microstructure and abrasion wear resistance of plasma sprayed Al_2O_3/TiO_2 composite coatings", *J. Manuf. Process.*, Vol. 10, pp. 1–5.

Morks, M. F. and Kobayashi, A., (2006), "Influence of gas flow rate on the microstructure and mechanical properties of hydroxyapatite coatings fabricated by gas tunnel type plasma spraying", *Surf. Coat. Technol.*, Vol. 201, pp. 2560–2566.

Morks, M. F., (2008), "Development of ZrO2/SiO2 bioinert ceramic coatings for biomedical application", *J. Mech. Behav. Biomed. Mater.*, Vol. 1, pp. 165–171.

Morks, M. F., Fahim, N. F. and Kobayashi, A., (2008), "Structure, mechanical performance and electrochemical characterization of plasma sprayed SiO_2/Ti-reinforced hydroxy-apatite biomedical coatings", *Appl. Surf. Sci.*, Vol. 255, pp. 3426–3433.

Morks, M. F., Kobayashi, A. and Fahim, N. F., (2007), "Abrasive wear behavior of sprayed hydroxyapatite coatings by gas tunnel type plasma spraying", *Wear*, Vol. 262, pp. 204–209.

Navarro, M. and Serra, T., (2016), "Biomimetic mineralization of ceramics and glasses", Elsevier BV. doi: 10.1016/B978-1-78242-338-6.00011-9.

Raimondi, M. T., Santambrogio, C., Pietrabissa, R., Raffelini, F. and Molfetta, L., (2001), "Improved mathematical model of the wear of the cup articular surface in hip joint prostheses and comparison with retrieved components", *Proc. Inst. Mech. Eng., Part H: J. Eng. Med.*, Vol. 215, pp. 377–390.

Riues, J., Rabbe, L. M. and Comrade, P., (1995), "Fretting wear corrosion of surgical implant alloys: effect of ion implantation and ion nitriding on fretting behavior of metals/PMMA contacts", In *Surface Modification Technologies VIII*, Eds. Sudarshan, M. and Jeandin, M., The Institute of Materials, pp. 43–52, London.

Saikko, V., Ahlroos, T. and Calonius, O., (2001), "A three-axis knee wear simulator with ball-on-flat contact", *Wear*, Vol. 249, pp. 310–315.

Sidhu, B. S., Singh, H., Puri, D. and Prakash, S., (2007), "Wear and oxidation behavior of shrouded plasma sprayed fly ash coatings", *Tribol. Int.*, Vol. 40, pp. 800–808.

Sun, L., Berndt, C. C., Gross, K. A. and Kucuk, A., (2001), "Material fundamentals and clinical performance of plasma-sprayed hydroxyapatite coatings: a review", *J. Biomed. Mater. Res.*, Vol. 58 (5), pp. 570–592.

Tercero, J. E., Namin, S., Lahiri, D., Balani, K., Tsoukias, N. and Agarwal, A., (2009), "Effect of carbon nanotube and aluminum oxide addition on plasma-sprayed hydroxyapatite coating's mechanical properties and biocompatibility", *Mater. Sci. Eng. C*, Vol. 29, pp. 2195–2202.

Van Citters, D. W., Kennedy, F. E., Currier, J. H., Collier, J. P. and Nichols, T. D., (2004), "A multi-station rolling/sliding tribotester for knee bearing materials", *ASME J. Tribol.*, Vol. 126, pp. 380–385.

Walker, P. S., Schneeweis, P., Murphy, S. and Nelson, P., (1987), "Strains and micromotions of press-fit femoral stem prostheses", *J. Biomech.*, Vol. 20, pp. 693–702.

Wang, A., Essner, A., Stark, C. and Dumbleton, J. H. (1999), "A biaxial line-contact wear machine for the evaluation of implant bearing materials for total knee joint replacement", *Wear*, Vol. 225–229, pp. 707–707.

Wang, Q., Shirong, G. and Dekun, Z., (2005), "Nano-mechanical properties and tribological behaviors of nanosized HA/partially-stabilized zirconia composites", *Wear*, Vol. 259, pp. 952–957.

Wang, Y., Khor, K. A. and Cheang, P., (1998), "Thermal spraying of functionally graded calcium phosphate coatings for biomedical implants", *J. Therm. Spray Technol.*, Vol. 7 (1), pp. 50–57.

Wroblewski, B. M., (1997), "Wear of high density polyethylene socket in total hip arthroplasty and its role in endosteal cavitation", *Proc. Inst. Mech. Eng., Part H, J. Eng. Med.*, Vol. 211 (H1), pp. 109–118.

2 Closed Form Solution for the Vibrational Analysis of Metal–Ceramic-Based Porous Functionally Graded Plate

Yogesh Kumar, Ankit Gupta, and Dheer Singh
Shiv Nadar University

CONTENTS

2.1 INTRODUCTION

Functionally graded (FG) materials are the superior tailored hybrid materials in which the properties of the material vary gradually in the preferred direction. Extensive literature has been published in the area of FG structures. In this context, Zhang et al. [1] presented an extensive review based on the buckling and vibrational response of the functionally graded plate (FGP). Gupta et al. [2–4] analyzed the vibrational behavior of a porous gradient plate using hybrid higher-order shear deformation theory (HSDT). Kanuet et al. [5] reviewed on smart functionally graded materials (FGMs) and analyzed buckling and vibration analysis based on fracture problems. Kurtaran et al. [6] studied the vibrational response of FGP and showed the effects of vibrating shape. Thai et al. [7] explored the vibrational and bending analysis of FGP using simple HSDT. Gupta and Talha [8] have studied and analyzed the effect of porosity on the structural response of a FGP using HSDT. They have reported that the fundamental frequency of the porous plate increases as the volume fraction index (VFI) increases. Significant changes have been reported in frequency for thin and thick plates. Shahverdi and Barati [9] have analyzed the free vibrational response of porous FGP. To investigate the vibrational response of a porous FGP on an elastic substrate, a nonlocal strain gradient elastic model has been developed. The authors have reported that the influence of various parameters such as nano-pores, nonlocal parameters, and gradient index have influence on fundamental frequencies of FGP.

Kumar et al. [10] proposed and carried out the geometrically non-linear analysis of the FGP using HSDT. Zenkour et al. [11] used the 3D elasticity solution for an Exponentially functionally graded plate (E-FGP). Birman V. et al. [12] proposed the research article on the modeling and investigation of FG structures. Reddy et al. [13] emphasized the comprehensive investigation of FGP. Benveniste et al. [14] suggested a new methodology for the use of Mori-Tanaka's principle FGP. Mori T et al. [15] reported average stresses in the average elastic energy in various materials.

Shariat et al. [16] presented the stability analysis of FGP with geometrical imperfection based on Classical plate theory (CPT). While using FSDT, Lanhe [17] derived the equilibrium and stability equations for a simply supported rectangular FGP with moderate thickness under thermal loads. Javaheri at al. [18,19] obtained the equilibrium and stability equation using variational approach method which is based on the CPT for perfect simply supported FGP. The authors have also derived the equilibrium equations under thermal loads based on the HSDT and power-law gradient rule. M Mohammadi et al. [20] developed a method to decouple the stability equation of moderately thick FGP using levy-type boundary conditions. An efficient and simply refined plate theory for studying buckling behavior of FGP using CPT was developed by Thai et al. [21]. Shariat et al. [22,23] introduced the rectangular FGP with or without geometric defects for mechanical and thermal analyses with the assumption of non-homogeneous mechanical properties varying through the plate thickness under the compression, tension, and thermal loading.

In this research work, the algebraic-based HSDT has been considered for the vibrational analysis of FGP. The power-law distribution and Hamilton's variational principle have been used to derive the governing equations. Governing equations are solved using Navier's method. The effects of various parameters like VFI and geometric parameter on the fundamental frequency of FGP have been reported in subsequent section.

The model of the porous FGP has been shown in Figure 2.1. The material properties of the porous FGP are shown in Table 2.1.

2.1.1 Mathematical Formulation

2.1.1.1 Displacement Field

The displacement field, i.e., $\bar{u}_i$ for $i = 1,2,3$, is the function of x, y, z, and t used in this work for the functional grade plate, reported by H.T. Thai [7], and expressed as the following equations:

$$\bar{u}_1 = \bar{u}_0 - z\frac{\partial \bar{w}_b}{\partial x} - \frac{4z^3}{3h^2}\left(\frac{\partial \bar{w}_s}{\partial x}\right)$$

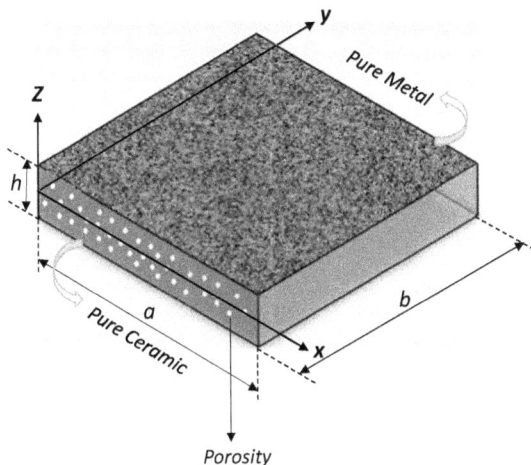

FIGURE 2.1   FG plates and its geometric coordinate.

TABLE 2.1
Material Properties of FGP [7]

S. No.	Properties	Ti-6AL-4V	Al$_2$O$_3$	ZrO$_2$
1	Young's modulus	105.7	380	200
2	Density ρ (kg/m³)	4429	3800	5700

$$\bar{u}_2 = \bar{v}_0 - z\frac{\partial \bar{w}_b}{\partial y} - \frac{4z^3}{3h^2}\left(\frac{\partial \bar{w}_s}{\partial y}\right)$$

$$\bar{u}_3 = \bar{w}_b(x,y,t) + \bar{w}_s(x,y,t) \qquad (2.1)$$

where $\bar{u}_0(x,y,t)$ and $\bar{v}_0(x,y,t)$ are the displacements in the x and y directions. $\bar{w}_b$ and $\bar{w}_s$ are the bending and shear components of the transverse displacement, respectively [7].

$$\varepsilon_{xx} = \frac{\partial \bar{u}_1}{\partial x} = \frac{\partial \bar{u}_0}{\partial x} - z\frac{\partial^2 \bar{w}_b}{\partial x^2} - \frac{4z^3}{3h^2}\left(\frac{\partial^2 \bar{w}_s}{\partial x^2}\right)$$

$$\varepsilon_{yy} = \frac{\partial \bar{u}_2}{\partial y} = \frac{\partial \bar{v}_0}{\partial y} - z\frac{\partial^2 \bar{w}_b}{\partial y^2} - \frac{4z^3}{3h^2}\left(\frac{\partial^2 \bar{w}_s}{\partial y^2}\right)$$

$$\varepsilon_{xy} = \frac{\partial \bar{u}_1}{\partial y} + \frac{\partial \bar{u}_2}{\partial x} = \frac{\partial \bar{u}_0}{\partial y} + \frac{\partial \bar{v}_0}{\partial x} - 2z \frac{\partial^2 \bar{w}_b}{\partial x \partial y} - \frac{8z^3}{3h^2} \left(\frac{\partial^2 \bar{w}_s}{\partial x \partial y} \right)$$

$$\varepsilon_{xz} = \frac{\partial \bar{u}_1}{\partial z} + \frac{\partial \bar{u}_3}{\partial x} = \left(1 - \frac{4z^2}{h^2} \right) \frac{\partial \bar{w}_s}{\partial x}$$

$$\varepsilon_{yz} = \frac{\partial \bar{u}_2}{\partial z} + \frac{\partial \bar{u}_3}{\partial y} = \left(1 - \frac{4z^2}{h^2} \right) \frac{\partial \bar{w}_s}{\partial y} \qquad (2.2)$$

2.1.1.2 Energy Equations

2.1.1.2.1 Strain Energy

Strain energy variation in FGP can be calculated by

$$\delta U = \int_A \int_{-\frac{h}{2}}^{+\frac{h}{2}} \left(\sigma_{xx} \delta \varepsilon_{xx} + \sigma_{yy} \delta \varepsilon_{yy} + \sigma_{xy} \delta \varepsilon_{xy} + \sigma_{xz} \delta \varepsilon_{xz} + \sigma_{yz} \delta \varepsilon_{yz} \right) dA \, dz \qquad (2.3)$$

where resultants of stresses are n, m, and q defined in the following equation:

$$\left(n_{xx}, n_{yy}, n_{xy} \right) = \int_{-\frac{h}{2}}^{+\frac{h}{2}} \left(\sigma_{xx}, \sigma_{yy}, \sigma_{xy} \right) dz$$

$$\left(m_{xx}^b, m_{yy}^b, m_{xy}^b \right) = \int_{-\frac{h}{2}}^{+\frac{h}{2}} \left(\sigma_{xx}, \sigma_{yy}, \sigma_{xy} \right) z \, dz$$

$$\left(m_{xx}^s, m_{yy}^s, m_{xy}^s \right) = \int_{-\frac{h}{2}}^{+\frac{h}{2}} \left(\sigma_{xx}, \sigma_{yy}, \sigma_{xy} \right) \frac{4z^3}{3h^2} dz$$

$$q_{xy} = \int_{-\frac{h}{2}}^{+\frac{h}{2}} \left(1 - \frac{4z^2}{h^2} \right) \sigma_{xy} \, dz$$

$$q_{yz} = \int_{-\frac{h}{2}}^{+\frac{h}{2}} \left(1 - \frac{4z^2}{h^2} \right) \sigma_{yx} \, dz \qquad (2.4)$$

- The work done δV is expressed in the following equation:

$$\delta V = - \int_A p \delta (\bar{w}_b + \bar{w}_s) dA \qquad (2.5)$$

where p is the transverse load.

- The kinetic energy δK is expressed in the following equation:

$$\delta K = \int_V (\dot{u}_1 \delta \dot{u}_1 + \dot{u}_2 \delta \dot{u}_2 + \dot{u}_3 \delta \dot{u}_3) \rho(z) dA dz \tag{2.6a}$$

$$\delta K = \int_A \left\{ \begin{array}{l} T_0[\dot{u}\delta\dot{u} + \dot{v}\delta\dot{v} + (\dot{w}_b + \dot{w}_s)\delta(\dot{w}_b + \dot{w}_s)] \\[2mm] -T_1\left[\dot{u}\dfrac{\partial \delta \dot{w}_b}{\partial x} + \dfrac{\partial \dot{w}_b}{\partial x}\delta\dot{u} + \dot{v}\dfrac{\partial \delta \dot{w}_b}{\partial y} + \dfrac{\partial \dot{w}_b}{\partial y}\delta\dot{v}\right] \\[4mm] +T_2\left[\dfrac{\partial \dot{w}_b}{\partial x} \times \dfrac{\partial \delta \dot{w}_b}{\partial x} + \dfrac{\partial \dot{w}_b}{\partial y} \times \dfrac{\partial \delta \dot{w}_b}{\partial y}\right] \\[4mm] +cT_3\left[\dot{u}\dfrac{\partial \delta \dot{w}_s}{\partial x} + \dfrac{\partial \dot{w}_s}{\partial x}\delta\dot{u} + \dot{v}\dfrac{\partial \delta \dot{w}_s}{\partial y} + \dfrac{\partial \dot{w}_s}{\partial y}\delta\dot{v}\right] \\[4mm] +cT_4\left[\begin{array}{l} \dfrac{\partial \dot{w}_b}{\partial x} \times \dfrac{\partial \delta \dot{w}_s}{\partial x} + \dfrac{\partial \dot{w}_s}{\partial x} \times \dfrac{\partial \delta \dot{w}_b}{\partial x} + \dfrac{\partial \dot{w}_b}{\partial y} \times \dfrac{\partial \delta \dot{w}_s}{\partial y} \\[3mm] +\dfrac{\partial \dot{w}_s}{\partial y} \times \dfrac{\partial \delta \dot{w}_b}{\partial y} \end{array} \right] \\[6mm] +c^2 T_6\left[\dfrac{\partial \dot{w}_s}{\partial x} \times \dfrac{\partial \delta \dot{w}_s}{\partial x} + \dfrac{\partial \dot{w}_s}{\partial y} \times \dfrac{\partial \delta \dot{w}_s}{\partial y}\right] \end{array} \right\} dA \tag{2.6b}$$

In Equation 2.6b, **dot** superscript represents the differentiation with respect to time (t), where mass density is $\rho(z)$, $c = \dfrac{4}{3h^2}$ and T_i represents mass inertia, given in the following equation:

$$(T_i) = \int_{-\frac{h}{2}}^{+\frac{h}{2}} (z^i) \rho(z) dz \quad \text{for } i = 0,1,2,3,4 \text{ and } 6 \tag{2.7}$$

2.1.1.2.2 Hamilton's Principle
Appling Hamilton's principle in the FGM plate,

$$0 = \int_0^T (\delta U + \delta V - \delta K) dt \tag{2.8}$$

When substituting all the value of δU, δV, and δK in Equation 2.8, we get

$$
\int_A \left\{
\begin{aligned}
&\left(n_{xx} \times \frac{\partial \delta \bar{u}_0}{\partial x} - m_{xx}^b \times \frac{\partial^2 \delta \bar{w}_b}{\partial x^2} - m_{xx}^s \times \frac{\partial^2 \delta \bar{w}_s}{\partial x^2} \right) \\[4pt]
&+ \left(n_{yy} \times \frac{\partial \delta \bar{v}_0}{\partial y} - m_{yy}^b \times \frac{\partial^2 \delta \bar{w}_b}{\partial y^2} - m_{yy}^s \times \frac{\partial^2 \delta \bar{w}_s}{\partial y^2} \right) \\[4pt]
&+ \left(\begin{aligned} & n_{xy}\left(\frac{\partial \delta \bar{u}_0}{\partial y} + \frac{\partial \delta \bar{v}_0}{\partial x} \right) - 2m_{xy}^b \times \frac{\partial^2 \delta \bar{w}_b}{\partial x \partial y} \\[4pt] & -2m_{xy}^s \times \frac{\partial^2 \delta \bar{w}_s}{\partial x \partial y} \end{aligned} \right) \\[4pt]
&+ \left(q_{xz} \times \frac{\partial \delta \bar{w}_s}{\partial x} \right) + \left(q_{yz} \times \frac{\partial \delta \bar{w}_s}{\partial y} \right) \\[4pt]
&- p\delta(\bar{w}_b + \bar{w}_s) \\[4pt]
&- T_0[\dot{\bar{u}}\delta\dot{\bar{u}} + \dot{\bar{v}}\delta\dot{\bar{v}} + (\dot{\bar{w}}_b + \dot{\bar{w}}_s)\delta(\dot{\bar{w}}_b + \dot{\bar{w}}_s)] \\[4pt]
&+ T_1\left[\dot{\bar{u}}\frac{\partial \delta \dot{\bar{w}}_b}{\partial x} + \frac{\partial \dot{\bar{w}}_b}{\partial x}\delta\dot{\bar{u}} + \dot{\bar{v}}\frac{\partial \delta \dot{\bar{w}}_b}{\partial y} + \frac{\partial \dot{\bar{w}}_b}{\partial y}\delta\dot{\bar{v}} \right] \\[4pt]
&- T_2\left[\frac{\partial \dot{\bar{w}}_b}{\partial x} \times \frac{\partial \delta \dot{\bar{w}}_b}{\partial x} + \frac{\partial \dot{\bar{w}}_b}{\partial y} \times \frac{\partial \delta \dot{\bar{w}}_b}{\partial y} \right] \\[4pt]
&+ cT_3\left[\dot{\bar{u}}\frac{\partial \delta \dot{\bar{w}}_s}{\partial x} + \frac{\partial \dot{\bar{w}}_s}{\partial x}\delta\dot{\bar{u}} + \dot{\bar{v}}\frac{\partial \delta \dot{\bar{w}}_s}{\partial y} + \frac{\partial \dot{\bar{w}}_s}{\partial y}\delta\dot{\bar{v}} \right] \\[4pt]
&- cT_4\left[\begin{aligned} & \frac{\partial \dot{\bar{w}}_b}{\partial x} \times \frac{\partial \delta \dot{\bar{w}}_s}{\partial x} + \frac{\partial \dot{\bar{w}}_s}{\partial x} \times \frac{\partial \delta \dot{\bar{w}}_b}{\partial x} + \frac{\partial \dot{\bar{w}}_b}{\partial y} \\[4pt] & \times \frac{\partial \delta \dot{\bar{w}}_s}{\partial y} + \frac{\partial \dot{\bar{w}}_s}{\partial y} \times \frac{\partial \delta \dot{\bar{w}}_b}{\partial y} \end{aligned} \right] \\[4pt]
&- c^2 T_6\left[\frac{\partial \dot{\bar{w}}_s}{\partial x} \times \frac{\partial \delta \dot{\bar{w}}_s}{\partial x} + \frac{\partial \dot{\bar{w}}_s}{\partial y} \times \frac{\partial \delta \dot{\bar{w}}_s}{\partial y} \right]
\end{aligned}
\right\} dA = 0 \qquad (2.9)
$$

By integrating Equation 2.9 by parts and then separating the coefficients of $\delta\bar{u}$, $\delta\bar{v}$, $\delta\bar{w}_b$, and $\delta\bar{w}_s$, the governing equations will be formed:

$$\delta \bar{u} = \left(\frac{\partial n_{xx}}{\partial x} + \frac{\partial n_{xy}}{\partial y} \right) = T_0 \ddot{u} + T_1 \ddot{u} \frac{\partial \ddot{w}_b}{\partial x} - cT_3 \frac{\partial \ddot{w}_s}{\partial x} \tag{2.10a}$$

$$\delta \bar{v} = \left(\frac{\partial n_{xy}}{\partial x} + \frac{\partial n_{y}}{\partial y} \right) = T_0 \ddot{v} - T_1 \frac{\partial \ddot{w}_b}{\partial y} - cT_3 \frac{\partial \ddot{w}_s}{\partial y} \tag{2.10b}$$

$$\delta \bar{w}_b = \left(\frac{\partial^2 m_{xx}^b}{\partial x^2} + \frac{\partial^2 m_{yy}^b}{\partial y^2} + 2\frac{\partial^2 m_{xy}^b}{\partial x \partial y} + p \right)$$

$$= T_0(\ddot{w}_b + \ddot{w}_s) + T_1 \left(\frac{\partial \ddot{u}}{\partial x} + \frac{\partial \ddot{v}}{\partial y} \right) - T_2 \nabla^2 \ddot{w}_b - cT_4 \nabla^2 \ddot{w}_s \tag{2.10c}$$

$$\delta \bar{w}_s = \left(\frac{\partial^2 m_{xx}^s}{\partial x^2} + \frac{\partial^2 m_{yy}^s}{\partial y^2} + 2\frac{\partial^2 m_{xy}^s}{\partial x \partial y} + \frac{\partial^2 q_{xz}}{\partial x} + \frac{\partial^2 q_{yz}}{\partial y} + p \right)$$

$$= T_0(\ddot{w}_b + \ddot{w}_s) + cT_3 \left(\frac{\partial \ddot{u}}{\partial x} + \frac{\partial \ddot{v}}{\partial y} \right) - cT_4 \nabla^2 \ddot{w}_b - c^2 T_6 \nabla^2 \ddot{w}_s \tag{2.10d}$$

2.1.1.3 Constitutive Relation

Power-law distribution has been used for the calculation of Young's modulus $E(z)$ shown in Equation 2.11. The porosity volume fraction (PVF) parameter has also included to observe the effect of even porosity distribution in porous FGP. E_m and E_c are Young's modulus of the metal and ceramic, respectively, and ξ and P are PVF and VFI, respectively [7].

$$E(z) = E_m + (E_c - E_m)(0.5 + z/h)^P - \frac{\xi}{2}(E_c + E_m) \tag{2.11}$$

$$
\begin{Bmatrix} \sigma_{xx} \\ \sigma_{yy} \\ \sigma_{xy} \\ \sigma_{xz} \\ \sigma_{yz} \end{Bmatrix} = \frac{E(z)}{1-v}
\begin{vmatrix}
1 & v & 0 & 0 & 0 \\
v & 1 & 0 & 0 & 0 \\
0 & 0 & \frac{1-v}{2} & 0 & 0 \\
0 & 0 & 0 & \frac{1-v}{2} & 0 \\
0 & 0 & 0 & 0 & \frac{1-v}{2}
\end{vmatrix}
\begin{Bmatrix} \varepsilon_{xx} \\ \varepsilon_{yy} \\ \varepsilon_{xy} \\ \varepsilon_{xz} \\ \varepsilon_{yz} \end{Bmatrix} \tag{2.12}
$$

Putting Equation 2.2 into the Equation 2.12 and obtained results into the Equation 2.5, we get

$$\begin{Bmatrix} n_{xx} \\ n_{yy} \\ n_{xy} \end{Bmatrix} = C_0 \begin{bmatrix} 1 & v & 0 \\ v & 1 & 0 \\ 0 & 0 & \dfrac{1-v}{2} \end{bmatrix} \begin{Bmatrix} \dfrac{\partial \bar{u}_0}{\partial x} \\ \dfrac{\partial \bar{v}_0}{\partial x} \\ \dfrac{\partial \bar{u}_0}{\partial x} + \dfrac{\partial \bar{v}_0}{\partial x} \end{Bmatrix} + C_1 \begin{bmatrix} 1 & v & 0 \\ v & 1 & 0 \\ 0 & 0 & \dfrac{1-v}{2} \end{bmatrix} \begin{Bmatrix} -\dfrac{\partial^2 \bar{w}_b}{\partial x^2} \\ -\dfrac{\partial^2 \bar{w}_b}{\partial y^2} \\ -2\dfrac{\partial^2 \bar{w}_b}{\partial x \partial y} \end{Bmatrix}$$

$$+ cC_3 \begin{bmatrix} 1 & v & 0 \\ v & 1 & 0 \\ 0 & 0 & \dfrac{1-v}{2} \end{bmatrix} \begin{Bmatrix} -\dfrac{\partial^2 \bar{w}_s}{\partial x^2} \\ -\dfrac{\partial^2 \bar{w}_s}{\partial y^2} \\ -2\dfrac{\partial^2 \bar{w}_s}{\partial x \partial y} \end{Bmatrix} \qquad (2.13)$$

$$\begin{Bmatrix} m_{xx}^b \\ m_{yy}^b \\ m_{xy}^b \end{Bmatrix} = C_1 \begin{bmatrix} 1 & v & 0 \\ v & 1 & 0 \\ 0 & 0 & \dfrac{1-v}{2} \end{bmatrix} \begin{Bmatrix} \dfrac{\partial \bar{u}_0}{\partial x} \\ \dfrac{\partial \bar{v}_0}{\partial x} \\ \dfrac{\partial \bar{u}_0}{\partial x} + \dfrac{\partial \bar{v}_0}{\partial x} \end{Bmatrix} + C_2 \begin{bmatrix} 1 & v & 0 \\ v & 1 & 0 \\ 0 & 0 & \dfrac{1-v}{2} \end{bmatrix} \begin{Bmatrix} -\dfrac{\partial^2 \bar{w}_b}{\partial x^2} \\ -\dfrac{\partial^2 \bar{w}_b}{\partial y^2} \\ -2\dfrac{\partial^2 \bar{w}_b}{\partial x \partial y} \end{Bmatrix}$$

$$+ cC_4 \begin{bmatrix} 1 & v & 0 \\ v & 1 & 0 \\ 0 & 0 & \dfrac{1-v}{2} \end{bmatrix} \begin{Bmatrix} -\dfrac{\partial^2 \bar{w}_s}{\partial x^2} \\ -\dfrac{\partial^2 \bar{w}_s}{\partial y^2} \\ -2\dfrac{\partial^2 \bar{w}_s}{\partial x \partial y} \end{Bmatrix} \qquad (2.14)$$

$$\begin{Bmatrix} m_{xx}^s \\ m_{yy}^s \\ m_{xy}^s \end{Bmatrix} = cC_3 \begin{bmatrix} 1 & v & 0 \\ v & 1 & 0 \\ 0 & 0 & \dfrac{1-v}{2} \end{bmatrix} \begin{Bmatrix} \dfrac{\partial \bar{u}_0}{\partial x} \\ \dfrac{\partial \bar{v}_0}{\partial x} \\ \dfrac{\partial \bar{u}_0}{\partial x} + \dfrac{\partial \bar{v}_0}{\partial x} \end{Bmatrix} + cC_4 \begin{bmatrix} 1 & v & 0 \\ v & 1 & 0 \\ 0 & 0 & \dfrac{1-v}{2} \end{bmatrix} \begin{Bmatrix} -\dfrac{\partial^2 \bar{w}_b}{\partial x^2} \\ -\dfrac{\partial^2 \bar{w}_b}{\partial y^2} \\ -2\dfrac{\partial^2 \bar{w}_b}{\partial x \partial y} \end{Bmatrix}$$

$$+ c^2 C_5 \begin{bmatrix} 1 & v & 0 \\ v & 1 & 0 \\ 0 & 0 & \dfrac{1-v}{2} \end{bmatrix} \begin{Bmatrix} -\dfrac{\partial^2 \bar{w}_s}{\partial x^2} \\ -\dfrac{\partial^2 \bar{w}_s}{\partial y^2} \\ -2\dfrac{\partial^2 \bar{w}_s}{\partial x \partial y} \end{Bmatrix} \qquad (2.15)$$

$$\begin{Bmatrix} q_{xz} \\ q_{yz} \end{Bmatrix} = A^s \begin{bmatrix} 1 & 0 \\ 0 & 1 \end{bmatrix} \begin{Bmatrix} \dfrac{\partial \bar{w}_s}{\partial x} \\ \dfrac{\partial \bar{w}_s}{\partial y} \end{Bmatrix} \qquad (2.16)$$

where C_i and A^s are the coefficients of stiffness defined by

$$(C_i) = \int_{-\frac{h}{2}}^{+\frac{h}{2}} (z^i) \frac{E(z)}{1-v^2} dz \quad \text{for } i = 0,1,2,3,4 \text{ and } 6 \tag{2.17a}$$

$$A^s = \int_{-\frac{h}{2}}^{+\frac{h}{2}} \left(1 - \frac{4z^2}{h^2}\right) \frac{E(z)}{2(1-v)} dz \tag{2.17b}$$

Equations 2.18a–2.18d are shown the equations of motion in terms of displacements:

$$C_o\left(\frac{\partial^2 \bar{u}_0}{\partial x^2} + \frac{1-v}{2}\frac{\partial^2 \bar{u}_0}{\partial y^2} + \frac{1+v}{2}\frac{\partial^2 \bar{v}_0}{\partial x \partial y}\right) - C_1 \nabla^2 \frac{\partial \bar{w}_b}{\partial x} - cC_3 \nabla^2 \frac{\partial \bar{w}_s}{\partial x}$$

$$= T_0 \ddot{\bar{u}} - T_1\left(\frac{\partial \ddot{\bar{w}}_b}{\partial x}\right) - cT_3\left(\frac{\partial \ddot{\bar{w}}_s}{\partial x}\right) \tag{2.18a}$$

$$C_o\left(\frac{\partial^2 \bar{v}_0}{\partial y^2} + \frac{1-v}{2}\frac{\partial^2 \bar{v}_0}{\partial x^2} + \frac{1+v}{2}\frac{\partial^2 \bar{u}_0}{\partial x \partial y}\right) - C_1 \nabla^2 \frac{\partial \bar{w}_b}{\partial y} - cC_3 \nabla^2 \frac{\partial \bar{w}_s}{\partial y}$$

$$= T_0 \ddot{\bar{v}} - T_1\left(\frac{\partial \ddot{\bar{w}}_b}{\partial y}\right) - cT_3\left(\frac{\partial \ddot{\bar{w}}_s}{\partial y}\right) \tag{2.18b}$$

$$C_1 \nabla^2\left(\frac{\partial \bar{u}_0}{\partial x} + \frac{\partial \bar{v}_0}{\partial y}\right) - C_2 \nabla^4 \bar{w}_b - cC_4 \nabla^4 \bar{w}_s + p = T_0(\ddot{\bar{w}}_b + \ddot{\bar{w}}_s)$$

$$+ T_1\left(\frac{\partial \ddot{\bar{u}}_0}{\partial x} + \frac{\partial \ddot{\bar{v}}_0}{\partial y}\right) - T_2 \nabla^2 \ddot{\bar{w}}_b - cT_4 \nabla^2 \ddot{\bar{w}}_s \tag{2.18c}$$

$$cC_3 \nabla^2\left(\frac{\partial \bar{u}_0}{\partial x} + \frac{\partial \bar{v}_0}{\partial y}\right) - cC_4 \nabla^4 \bar{w}_b - c^2 C_5 \nabla^4 \bar{w}_s + A^s \nabla^2 \bar{w}_s + p$$

$$= T_0(\ddot{\bar{w}}_b + \ddot{\bar{w}}_s) + cT_3\left(\frac{\partial \ddot{\bar{u}}_0}{\partial x} + \frac{\partial \ddot{\bar{v}}_0}{\partial y}\right) - cT_4 \nabla^2 \ddot{\bar{w}}_b - c^2 T_6 \nabla^2 \ddot{\bar{w}}_s \tag{2.18d}$$

2.1.1.4 Analytical Solution

Applying the Navier solution to the functional-grade plate, we get

$$\bar{u}_0 = \sum_{m=1}^{\infty}\sum_{n=1}^{\infty} \bar{U}_{mn} e^{i\omega t} \cos \alpha x \sin \beta y$$

$$\bar{v}_0 = \sum_{m=1}^{\infty}\sum_{n=1}^{\infty} \bar{V}_{mn} e^{i\omega t} \sin \alpha x \cos \beta y$$

$$\bar{w}_b = \sum_{m=1}^{\infty}\sum_{n=1}^{\infty} \bar{W}_{bmn} e^{i\omega t} \sin \alpha x \sin \beta y$$

$$\bar{w}_s = \sum_{m=1}^{\infty}\sum_{n=1}^{\infty} \bar{W}_{smn} e^{i\omega t} \sin \alpha x \sin \beta y$$

(2.19a)

where $\alpha = \dfrac{m\pi}{a}$, $\beta = \dfrac{n\pi}{b}$ and $i = \sqrt{-1}$.

$\left(\bar{U}_{mn}, \bar{V}_{mn}, \bar{W}_{bmn} \text{ and } \bar{W}_{smn}\right)$ are coefficients, where ω is the angular frequency and the p is the transverse load expanded in the double Fourier sine series, which is given by

$$p(x,y) = \sum_{m=1}^{\infty}\sum_{n=1}^{\infty} q_{mn} \sin \alpha x \sin \beta y$$

(2.19b)

The coefficient q_{mn} can be represented as

$$q_{mn} = \frac{4}{ab} \int_0^a \int_0^b p(x,y) \sin \alpha x \sin \beta y$$

(2.19c)

Substitute Equations 2.19a and 2.19b in Equation 2.18; then the stiffness matrix coefficients and mass matrix coefficients are calculated from the above equations are shown in the following equation:

$$\left(\begin{bmatrix} \hat{K}_{11} & \hat{K}_{12} & \hat{K}_{13} & \hat{K}_{14} \\ \hat{K}_{12} & \hat{K}_{22} & \hat{K}_{23} & \hat{K}_{24} \\ \hat{K}_{13} & \hat{K}_{23} & \hat{K}_{33} & \hat{K}_{34} \\ \hat{K}_{14} & \hat{K}_{24} & \hat{K}_{34} & \hat{K}_{44} \end{bmatrix} - \omega^2 \begin{bmatrix} M_{11} & 0 & M_{13} & M_{14} \\ 0 & M_{22} & M_{23} & M_{24} \\ M_{13} & M_{23} & M_{33} & M_{34} \\ M_{14} & M_{24} & M_{34} & M_{44} \end{bmatrix} \right) \begin{Bmatrix} \bar{U}_{mn} \\ \bar{V}_{mn} \\ \bar{W}_{bmn} \\ \bar{W}_{smn} \end{Bmatrix} = \begin{Bmatrix} 0 \\ 0 \\ q_{mn} \\ q_{mn} \end{Bmatrix}$$

(2.20)

$$\hat{K}_{11} = C_0\alpha^2 + \frac{1-v}{2}C_0\beta^2, \ \hat{K}_{12} = \frac{1+v}{2}C_0\alpha\beta, \ \hat{K}_{13} = -C_1\alpha\left(\alpha^2 + \beta^2\right)$$

$$\hat{K}_{14} = -cC_3\alpha\left(\alpha^2 + \beta^2\right), \ \hat{K}_{22} = \frac{1-v}{2}C_0\alpha^2 + C_0\beta^2, \ \hat{K}_{23} = -C_1\beta\left(\alpha^2 + \beta^2\right)$$

(2.21)

$$\hat{K}_{24} = -cC_3\beta\left(\alpha^2 + \beta^2\right), \hat{K}_{33} = C_2\left(\alpha^2 + \beta^2\right)^2, \ \hat{K}_{34} = cC_3\left(\alpha^2 + \beta^2\right)^2$$

$$\hat{K}_{44} = c^2 C_5\left(\alpha^2 + \beta^2\right)^2 + A^s\left(\alpha^2 + \beta^2\right)$$

$$M_{11} = M_{22} = I_0, M_{12} = 0, M_{13} = \alpha I_1, M_{14} = -\alpha c I_3, M_{23} = -\beta I_1, M_{24} = -\beta c I_3$$

(2.22)

$$M_{33} = I_0 + I_1\left(\alpha^2 + \beta^2\right), M_{34} = I_0 + cI_4\left(\alpha^2 + \beta^2\right), M_{44} = I_0 + c^2 I_6\left(\alpha^2 + \beta^2\right)$$

2.2 RESULTS AND DISCUSSION

The fundamental frequency values are being validated in this research work and compared with the different non-dimensional frequencies (NDFs) of a FGP governed by various HSDTs. The non-dimensional parameter $\hat{\omega} = \omega h \sqrt{\rho_c / E_c}$ is used for the analysis, and the different properties of metal and ceramic materials are shown in Table 2.1.

Table 2.2 shows that this work has been validated using HTDTs used by HT Thai [7]. These results for the resonant frequency of Al/Al_2O_3 of thin and thick square FGP are in good agreement with referred results. The validation of the Al/Al_2O_3 FGP used for different side-to-thickness ratios (a/h) and power-law indexes and various modes of vibration have been carried out. It has been found that most of the values of non-dimensional frequencies are absolutely validated and some show the error of approximately 1% in each mode of vibration.

The non-dimensional fundamental frequency analysis of materials Ti-$6AL$-$4V/$ ZrO_2, i.e., PM1, and Ti-$6AL$-$4V/Al_2O_3$, i.e., PM2, for thin and thick square FGP has been carried out for several values of side-to-thickness ratios and power-law indexes, and the results reported in both Tables 2.3 and 2.4 show that the NDF value is decreasing as the values of the VFI increase, i.e., by decreasing the VFI in the FG plate. Furthermore, as the thickness ratio increased, the NDF of the plate will decrease.

The NDF of PM1 and PM2 for thin FGP has been calculated for several values of aspect ratios (a/b) using power-law indexes. The analysis has been performed at fixed a/h, i.e., $a/h = 10$. Results are reported in both Tables 2.5 and 2.6. The NDF values are decreasing with the increasing values of VFI, i.e., decreasing the ceramic content in the FGP.

Figures 2.2–2.7 show the variation of frequency at different modes of vibration with varying VFI (P) for PM1 and PM2. The variation reduction in frequency has been found to be steep initially, but after VFI crosses ten, it becomes almost stable for all modes of frequency.

Tables 2.7 and 2.8 show the change in fundamental frequency with varying VFI, side-to-thickness ratios, and PVF for both metal and ceramic-based porous FGP.

TABLE 2.2
Result Validation Based on HSDT [7] and Present Work for Al/Al_2O_3

P		0		1		4		10	
		HSDT		HSDT		HSDT		HSDT	
a/h	Mode	[7]	Present	[7]	Present	[7]	Present	[7]	Present
5	1(1,1)	0.2113	0.2112	0.1631	0.1631	0.1378	0.1378	0.1301	0.1301
	2(1,2)	0.4623	0.4625	0.3607	0.3607	0.298	0.298	0.2771	0.2771
	3(2,2)	0.6688	0.6688	0.5254	0.5245	0.4284	0.4284	0.3948	0.3948
10	1(1,1)	0.0577	0.0577	0.0442	0.0442	0.0381	0.0381	0.0364	0.0364
	2(1,2)	0.1377	0.1376	0.1059	0.1053	0.0903	0.0903	0.0856	0.0866
	3(2,2)	0.2113	0.2113	0.1631	0.1631	0.1378	0.1388	0.1301	0.1304

TABLE 2.3

Non-dimensional Frequencies (ω) Using *PM2* Material

		P					
a/h	Mode	0	0.5	1	4	10	50
5	1(1,1)	0.2113	0.1726	0.1543	0.1294	0.1206	0.1092
	2(1,2)	0.4623	0.3799	0.3396	0.2795	0.259	0.2375
	3(2,2)	0.6688	0.5517	0.4933	0.4015	0.3706	0.3423
10	1(1,1)	0.0577	0.0469	0.042	0.0357	0.0335	0.0300
	2(1,2)	0.1376	0.1122	0.1004	0.0847	0.0792	0.0713
	3(2,2)	0.2113	0.1726	0.1543	0.1294	0.1206	0.1092
20	1(1,1)	0.0148	0.012	0.0108	0.0092	0.0086	0.0077
	2(1,2)	0.0365	0.0297	0.0266	0.0227	0.0212	0.0190
	3(2,2)	0.0577	0.0469	0.042	0.0357	0.0335	0.0300

TABLE 2.4

Non-dimensional Frequencies (ω) Using *PM1* Material

		P					
a/h	Mode	0	0.5	1	4	10	50
5	1(1,1)	0.2113	0.1985	0.1933	0.1872	0.1835	0.1773
	2(1,2)	0.4623	0.4358	0.4239	0.4065	0.3893	0.387
	3(2,2)	0.6688	0.6319	0.6142	0.5854	0.5735	0.5592
10	1(1,1)	0.0577	0.0541	0.0527	0.0515	0.0505	0.0485
	2(1,2)	0.1376	0.1292	0.1258	0.1223	0.12	0.1156
	3(2,2)	0.2113	0.1985	0.1933	0.1872	0.1835	0.1773
20	1(1,1)	0.0148	0.0139	0.0135	0.0132	0.013	0.0125
	2(1,2)	0.0365	0.0342	0.0333	0.0326	0.032	0.0307
	3(2,2)	0.0577	0.0541	0.0527	0.0515	0.0505	0.0485

TABLE 2.5

Non-dimensional Frequencies (ω) Using *PM2*

		P					
a/b	Mode	0	0.5	1	4	10	50
1	1(1,1)	0.0577	0.0469	0.042	0.0357	0.0335	0.03
	2(1,2)	0.1376	0.1122	0.1004	0.0847	0.0792	0.0713
	3(2,2)	0.2113	0.1726	0.1543	0.1294	0.1206	0.1092
1.5	1(1,1)	0.0919	0.0748	0.0669	0.0568	0.0531	0.0477
	2(1,2)	0.2574	0.2106	0.1883	0.1572	0.1464	0.1329
	3(2,2)	0.323	0.2646	0.2366	0.1965	0.1827	0.1665
2	1(1,1)	0.1376	0.1122	0.1004	0.0847	0.0792	0.0713
	2(1,2)	0.4047	0.3321	0.2969	0.2453	0.2276	0.2081
	3(2,2)	0.4623	0.3799	0.3396	0.2795	0.259	0.2375

TABLE 2.6
Non-dimensional Frequencies (ω) Using *PM1*

a/b	Mode	P					
		0	0.5	1	4	10	50
1	1(1,1)	0.0577	0.0541	0.0527	0.0515	0.0505	0.0485
	2(1,2)	0.1376	0.1292	0.1258	0.1223	0.12	0.1156
	3(2,2)	0.2113	0.1985	0.1933	0.1872	0.1835	0.1773
1.5	1(1,1)	0.0919	0.0862	0.084	0.0818	0.0802	0.0772
	2(1,2)	0.2574	0.2421	0.2356	0.2277	0.2232	0.2159
	3(2,2)	0.323	0.304	0.2958	0.2851	0.2795	0.2708
2	1(1,1)	0.1376	0.1292	0.1258	0.1223	0.12	0.1156
	2(1,2)	0.4047	0.3813	0.3709	0.3564	0.3492	0.339
	3(2,2)	0.4623	0.4358	0.4239	0.4065	0.3893	0.387

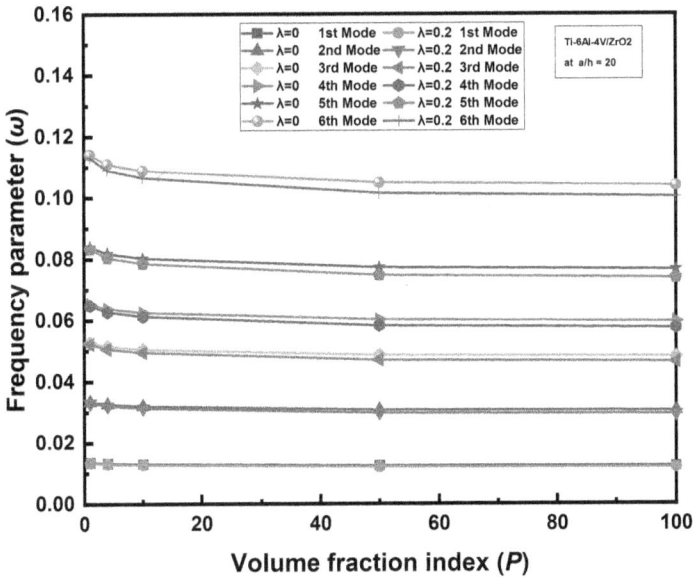

FIGURE 2.2 Higher modes vs. VFI (*P*) at *a/h* = 20 for PM1.

For PM2, the values of NDF decrease with increasing VFI and PVF. It has been observed that when the plate is changed from ceramic to metallic, the fundamental frequency decreases. The reduction in frequency parameter with increasing VFI up to 100 is approximately 33%, whereas with the increase in PVF up to 0.4, the reduction in frequency parameter found to be around 10%.

For PM1, a similar reduction percentage has also been observed in this case. The frequency parameter values have been reduced for the increase in VFI and PFI. The frequency parameter values have been found to be higher than PM2.

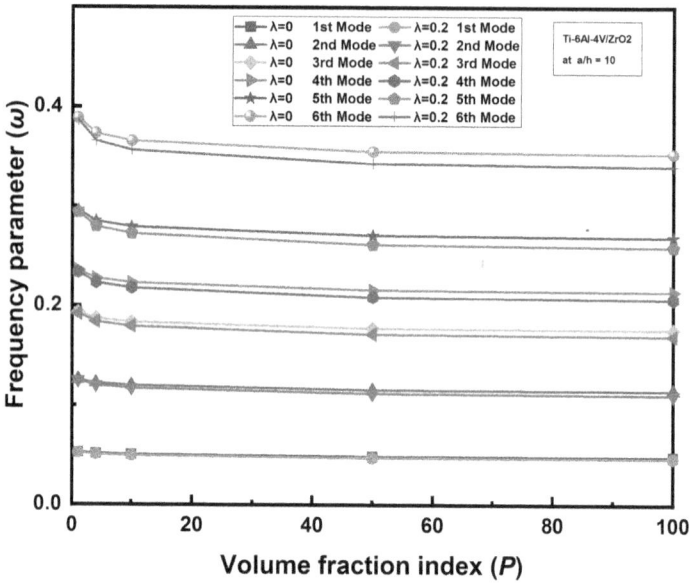

FIGURE 2.3   Higher modes vs. VFI (*P*) at *a/h*= 10 for PM1.

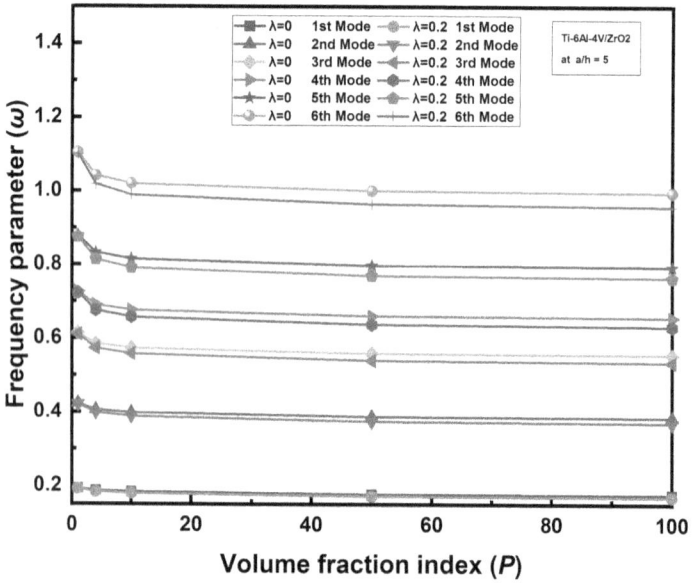

FIGURE 2.4   Higher modes vs. VFI (*P*) at *a/h*=5 for PM1.

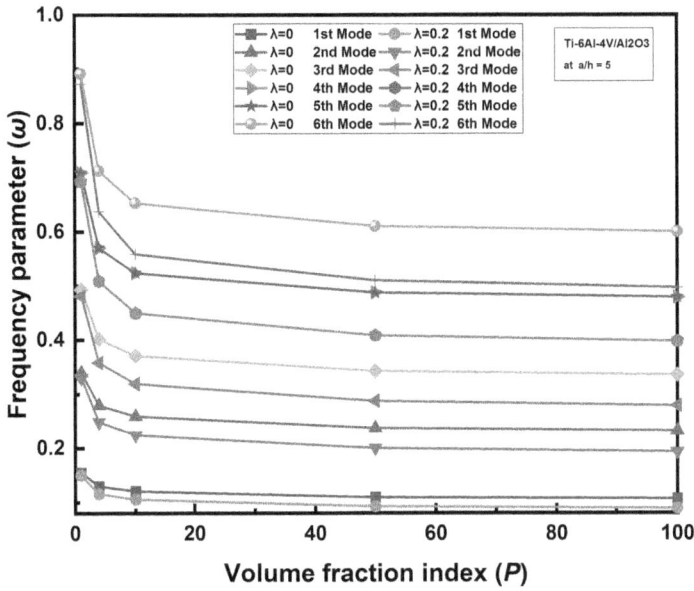

FIGURE 2.5   Higher modes vs. VFI (P) at a/h = 5 for PM2.

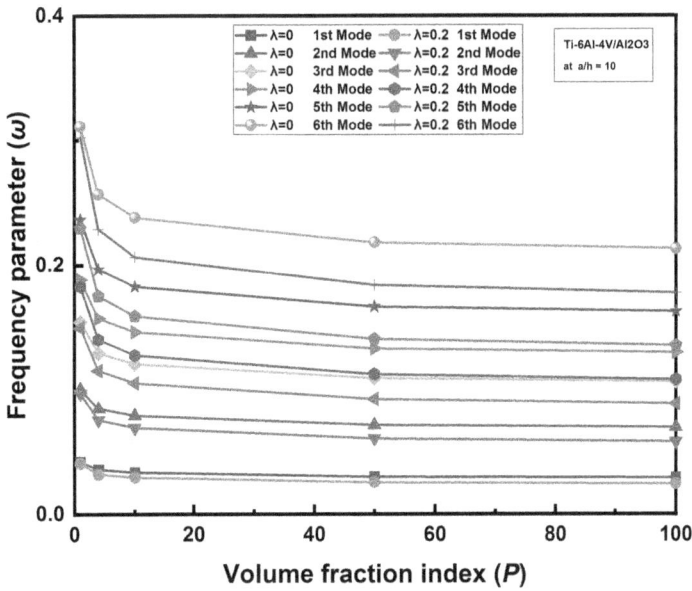

FIGURE 2.6   Higher modes vs. VFI (P) at a/h = 10 for PM2.

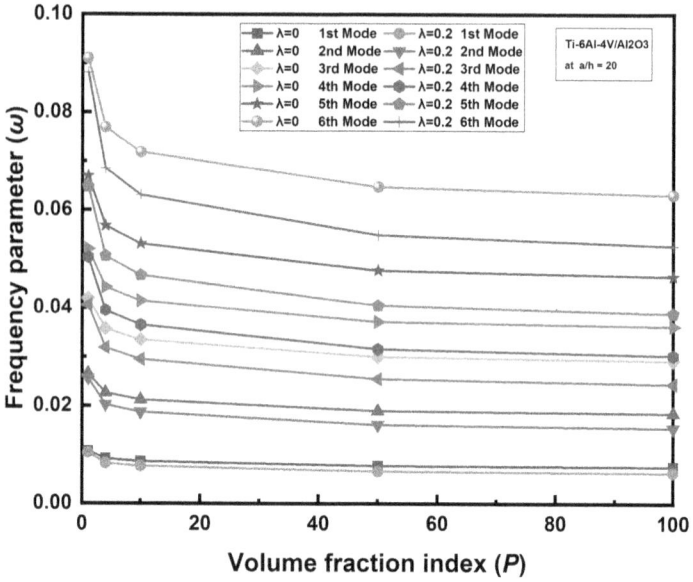

FIGURE 2.7 Higher modes vs. VFI (*P*) at *a/h*=5 for PM2.

TABLE 2.7
Frequency Parameter with Varying *P* and ξ at *a/b*=1 for PM2

PM2		VFI				
a/h	ξ	1	4	10	50	100
5	0	0.1543	0.1294	0.1206	0.1092	0.1065
	0.1	0.1524	0.1236	0.1143	0.1021	0.0990
	0.2	0.1496	0.1153	0.1054	0.0924	0.0888
	0.3	0.1453	0.1017	0.0910	0.0782	0.0739
	0.4	0.1385	0.0731	0.0589	0.0500	0.0463
10	0	0.0420	0.0357	0.0335	0.0300	0.0292
	0.1	0.0414	0.0342	0.0318	0.0281	0.0271
	0.2	0.0406	0.0319	0.0295	0.0255	0.0244
	0.3	0.0394	0.0281	0.0257	0.0217	0.0204
	0.4	0.0374	0.0199	0.0167	0.0143	0.0130
20	0	0.0108	0.0092	0.0086	0.0077	0.0075
	0.1	0.0106	0.0088	0.0082	0.0072	0.0070
	0.2	0.0104	0.0082	0.0076	0.0066	0.0063
	0.3	0.0101	0.0072	0.0067	0.0056	0.0052
	0.4	0.0096	0.0051	0.0043	0.0037	0.0034

TABLE 2.8

Frequency Parameter with Varying P and ξ at $a/b=1$ for PM1

PM1		VFI				
a/h	ξ	1	4	10	50	100
5	0	0.1933	0.1872	0.1835	0.1773	0.1759
	0.1	0.1927	0.1857	0.1817	0.1747	0.1731
	0.2	0.1918	0.1837	0.1793	0.1714	0.1695
	0.3	0.1905	0.1807	0.1759	0.1667	0.1645
	0.4	0.1884	0.1760	0.1707	0.1599	0.1571
10	0	0.0527	0.0515	0.0505	0.0485	0.0481
	0.1	0.0525	0.0511	0.0500	0.0478	0.0473
	0.2	0.0522	0.0506	0.0494	0.0469	0.0464
	0.3	0.0518	0.0498	0.0486	0.0457	0.0450
	0.4	0.0512	0.0486	0.0473	0.0439	0.0430
20	0	0.0135	0.0132	0.0130	0.0125	0.0123
	0.1	0.0135	0.0131	0.0129	0.0123	0.0121
	0.2	0.0134	0.0130	0.0127	0.0121	0.0119
	0.3	0.0133	0.0128	0.0125	0.0117	0.0116
	0.4	0.0131	0.0125	0.0122	0.0113	0.0110

2.3 CONCLUSIONS

In this study, the algebraic-based HSDT has been considered for the vibrational response of the porous FGP using the Navier solution. It has been found that the fundamental frequency decreases as the VFI increases. It is also concluded that the free vibrational frequency will decrease with an increase in the a/h ratio. The effect of VFI and side-to-thickness ratio has also been reported for rectangular plates in addition to square plates. Higher modes of vibration have also been reported in this analysis. It has been shown that the NDF of the plate increases with the increases in modes of vibration. The percentage difference at higher modes of vibration in porosity models is quite high in comparison with that at lower modes.

It is shown in the above results that as the FGP is changed from purely ceramic to purely metal, the natural frequency of the FGP decreases. The vibration analysis results given in this study are useful for the design of porous FGP.

REFERENCES

1. N. Zhang, T. Khan, H. Guo (2019) Functionally graded materials: an overview of stability, buckling, and free vibration analysis, *Adv. Mater. Sci. Eng.*, Article ID 1354150, 18.
2. A. Gupta, M. Talha (2018) Influence of porosity on the flexural and vibration response of gradient plate using non polynomial higher-order shear and normal deformation theory, *Int. J. Mech. Mater.*, 14, 277–296.

3. A. Gupta, M. Talha (2017) Large amplitude free flexural vibration analysis of finite element modeled FGM plates using new hyperbolic shear and normal deformation theory, *Aerosp. Sci. Technol.*, 67, 287–308.
4. A. Gupta, M. Talha (2015) Recent development in modeling and analysis of functionally graded materials and structures, *Prog. Aerosp. Sci.*, 79, 1–14.
5. N. J. Kanu, U. K. Vates (2017) Fracture problems, vibration, buckling, and bending analyses of functionally graded materials: a state-of-the-art review including smart FGMS, *Part. Sci. Tech.* ISSN: 0272-6351.
6. H. Kurtaran, (2014) Shape effect on free vibration of functionally graded plates, *Int. J. Eng. Appl. Sci.*, 6(4), 52–67.
7. H.-T. Thai, S.-E. Kim (2013) A simple higher-order shear deformation theory for bending and free vibration analysis of functionally graded plates, *Compos. Struct.*, 96, 165–173.
8. A. Gupta, M. Talha (Mach 2018) Influence of initial geometric imperfections and porosity on the stability of functionally graded material plates. *Mech. Based Des. Struct.*, 46, 693–711. https://doi.org/10.1080/15397734.2018.1449656.
9. L. F. Qian, R. C. Batra, L. M. Chen (2004) Static and dynamic deformations of thick functionally graded elastic plates by using higher-order shear and normal deformable plate theory and meshless local Petrov–Galerkin method. *Compos. Part B Eng.*, 35, 685–697. https://doi.org/10.1016/j.compositesb.2004.02.004.
10. J. S. Kumar, B. S. Reddy (2011) Geometrically nonlinear analysis of functionally graded material plates using higher order theory, *Int. J. Eng. Sci. Technol.*, 3(1), 279–288
11. A. M. Zenkour (2007) Benchmark trigonometric and 3-D elasticity solutions for an exponentially graded thick rectangular plate, *Arch. Appl. Mech.*, 77(4), 197–214.
12. V. Birman, L. W. Byrd (2007) Modeling and analysis of functionally graded materials and structures, *Appl. Mech. Rev.*, 60, 195.
13. J. N. Reddy (2000) Analysis of functionally graded plates, *Int. J. Numer. Methods Eng.*, 47(1–3), 663–684.
14. Y. Benveniste (1987) A new approach to the application of Mori–Tanaka's theory in composite materials, *Mech. Mater.*, 6(2), 147–157.
15. T. Mori, K. Tanaka (1973) Average stress in matrix and average elastic energy of materials with misfitting inclusions, *Acta Metall.*, 21(5), 571–574.
16. Shariat, S., Eslami, M. R. (2005) Buckling of functionally graded plates under in plane compressive loading based on the first order plate theory, *Proceeding of the Fifth International Conference on Composite Science and Technology*, American University of Sharjah, United Arab Emi.
17. W. Lanhe (2004) Thermal buckling of a simply supported moderately thick rectangular FGM plate, *Compos. Struct.*, 64(2), 211–218.
18. R. Javaheri, M. R. Eslami (2002) Buckling of functionally graded plates under in-plane compressive loading, *ZAMM Zeitschrift fur Angew. Math. und Mech.*, 82(4), 277–283.
19. R. Javaheri, M. R. Eslami (2002) Thermal buckling of functionally graded plates based on higher order theory, *J. Therm. Stress.*, 25(7), 603–625.
20. M. Mohammadi, A. R. Saidi, E. Jomehzadeh (2010) A novel analytical approach for the buckling analysis of moderately thick functionally graded rectangular plates with two simply-supported opposite edges, *Proc. Inst. Mech. Eng. Part C J. Mech. Eng. Sci.*, 224(9), 1831–1841.
21. H.-T. Thai (2012) An efficient and simple refined theory for buckling analysis of functionally graded plates, *Appl. Math. Model.*, 36, 1008–1022.
22. B. A. S. Shariat, M. R. Eslami (2006) Thermomechanical stability of imperfect functionally graded plates based on the third-order theory, *AIAA J.*, 44(12), 2929–2936.
23. B. A. S. Shariat, M. R. Eslami (2007) Buckling of thick functionally graded plates under mechanical and thermal loads, *Compos. Struct.*, 78(3), 433–439.

3 Functionally Graded Structures

Design and Analysis of Tailored Advanced Composites

Ankit Gupta
Shiv Nadar University

CONTENTS

3.1 INTRODUCTION

Functionally graded materials (FGMs) are advanced heterogeneous composite materials developed using two or more constituent materials with a continuously varying composition. These advanced materials possess extraordinary properties compared to homogeneous material made of similar constituents due to engineered gradients of composition in the preferred direction (Birman & Byrd, 2007; Jha, Kant, & Singh, 2013). FGMs have no abrupt change in material properties across the interface of dissimilar materials; therefore, it circumvents the problems such as large inter-laminar stresses, crack initiation, and de-lamination problems associated with conventional composite materials (Mojdehi & Darvizeh, 2011). The concept of FGMs was first introduced by Bever and Duwez (1972) and was further explored by the group of Japanese scientists in National Aerospace Laboratories in 1984 to use them as a thermal barrier material (Koizumi, 1997). In recent years, FGMs have proven its worth in various industries such as aerospace, mechanical, nuclear, and bioengineering due to their ability to produce materials with requisite properties.

As discussed earlier, FGMs are high-performance, microscopically heterogeneous tailored advanced composite materials. Such materials are generally fabricated using two or more constituent phases with the continual variation of volume fractions in the preferred orientation (Gupta & Talha, 2015). FGMs integrate the best chemical, mechanical, and thermal properties of two materials (ceramics, metal, polymers, etc.) into a single part with a smooth graded interface which makes it able to withstand in the extreme working environment. An uninterruptedly graded microstructure with typical metal/ceramic FGM is shown in Figure 3.1. Ceramic (zirconia in the figure) alone is thermally and chemically resistant with low impact toughness. Metal (Ni-alloy in the figure) alone is tough and strong enough, but its mechanical performance is sensitive to the thermal environment. On the other hand, metal-to-ceramic (Ni-alloys/ZrO_2) FGMs are able to sustain over 1000°C temperature gradient over a thickness of 10mm only. Such FGMs are capable of withstanding under extremely high thermal and chemical environments at the ceramic face while maintaining overall strength and resistance to brittle fracture due to the metallic content (Koizumi, 1997; Koizumi & Niino, 1995).

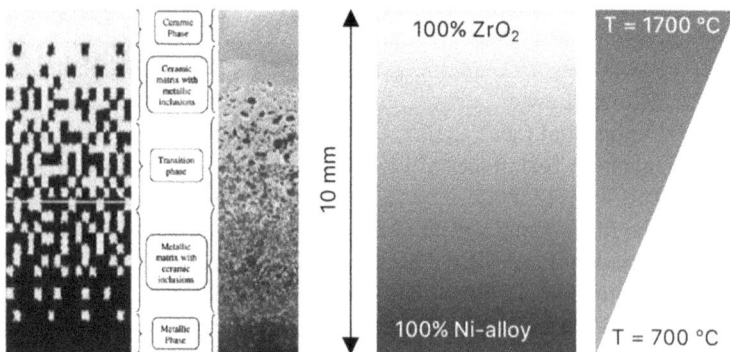

FIGURE 3.1 Gradation of microstructure with metal–ceramic FGMs (Gupta & Talha, 2015). (https://aerosint.com/home/how-to-make-cheap-scalable-multi-material-3d-printing-a-reality/.)

The concept of FGMs is not new to nature. For example, the human body possesses FGMs like materials in many parts such as bones, skin, or teeth (Lin, Li, Li, & Swain, 2010) as shown in Figure 3.2a. A tooth (more specifically dental crowns) is an example of the FGMs in the human body as it necessitates a high wear resistance outside (enamel), and a ductile inner structure for fatigue and brittleness (Figure 3.2b) (He & Swain, 2009). In addition to this, FGMs are also evident in plants and in animal tissues (Silva, Walters, & Paulino, 2006). Bamboo is also a good example of natural graded structure, composed of hierarchical fiber and parenchyma structures that evolve to resist wind loads (Figure 3.2 (c)). These bamboo structures exhibit outstanding properties, such as stiffness, strength, and fracture resistance. Some examples of natural FGMs are presented in Figure 3.2.

3.2   TYPES OF FUNCTIONALLY GRADED MATERIALS

The extraordinary properties and applications in diversified industries seized the interest of the scientist and researchers to tailor the different types of FGMs. The selection of the different type of FGMs depends on the applications for which it is being used. Broadly, the FGMs can be of three types: microstructural gradient FGMs (MG-FGMs), chemical composition gradient FGMs (CCG-FGMs), and porosity gradient-based FGMs (PG-FGMs) (Mahamood & Akinlabi, 2017).

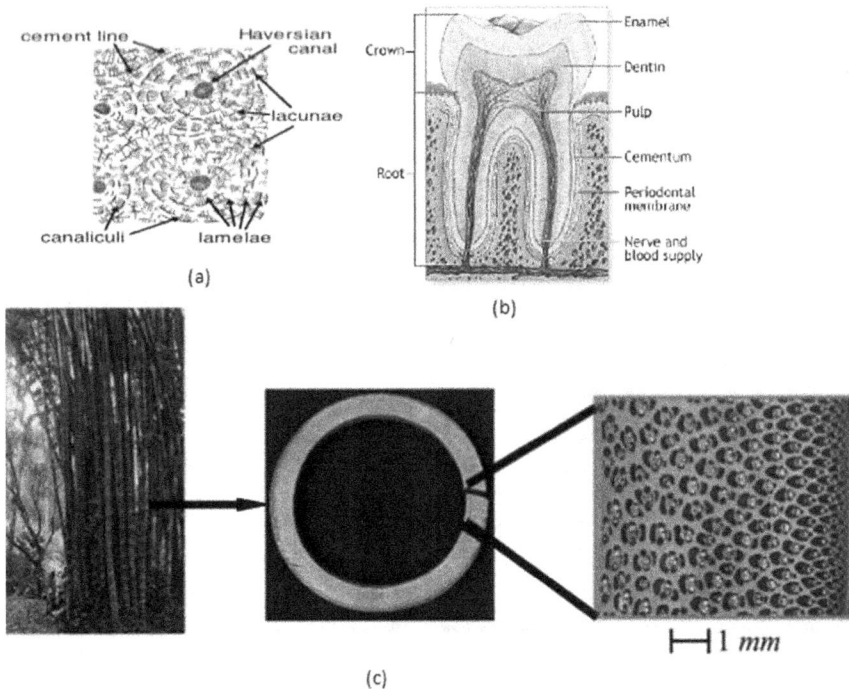

FIGURE 3.2   Natural functionally graded materials (Palmer, Newcomb, Kaltz, Spoerke, & Stupp, 2008; Silva et al., 2006).

In MG-FGMs, the microstructure of the constituents gradually changes to achieve the required properties. Such types of FGMs are generally used when very hard surface properties are required such as in Ballistic armors or in the defense sector. Such FGMs can be fabricated by using controlled heat treatment process (Mahamood & Akinlabi, 2017; Chenglin, Jingchuan, Zhongda, & Shidong, 1999; Zhu, Yin, & Lai, 1996). In CCG-FGMs, the chemical composition of the constituent continuously changes as per the spatial position in the material. CCG-FGMs can be further categorized into single-phase and multi-phase material. Such types of FGMs are generally manufactured by powder metallurgy where different chemical compounds and phases formed during the sintering process. These FGMs have wide applications in automobile, aerospace, and energy sectors. In PG-FGMs, porosity in the material is intentionally incorporated to have the gradual change of constituent volume fraction. The size and shape of the pore are considered according to the desired properties. Such types of FGMs have a very promising application in biomedical field where porous FGMs are very much required for the better combination of the implant and the surrounding tissues (Mahamood, & Akinlabi, 2015; Suk, Choi, Kim, Kim, Kwon, 2003).

Apart from the aforementioned types of FGMs, it can also be categorized according to the nature of gradient as shown in Figure 3.3 (Bharti, Gupta, & Gupta, 2013). These FGMs are fraction gradient type (Figure 3.3a), size gradient type (Figure 3.3b), orientation gradient type (Figure 3.3c), and shape gradient type (Figure 3.3d).

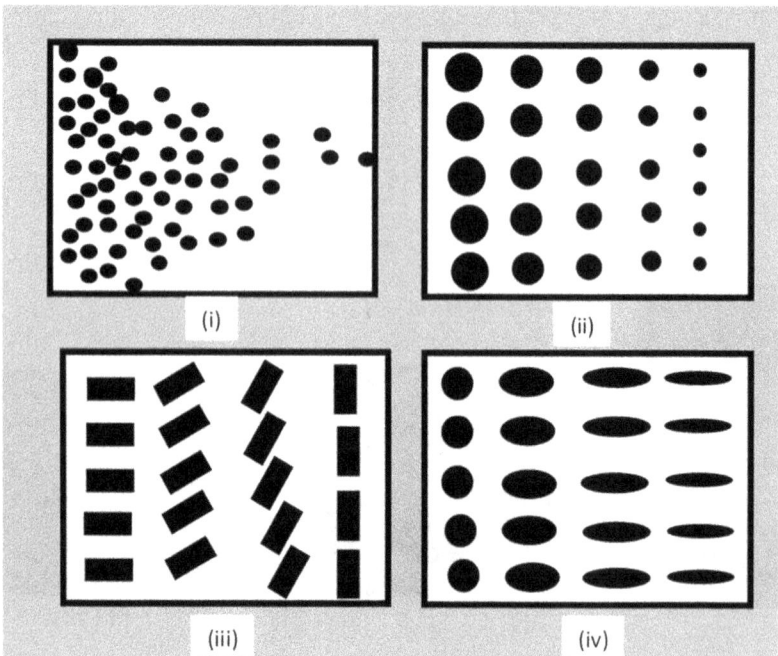

FIGURE 3.3 Different types of functionally graded materials: (a) fraction gradient type, (b) size gradient type, (c) orientation gradient type, and (d) shape gradient type.

3.3 PROCESSING TECHNIQUES OF FGMs

3.3.1 DRY POWDER PROCESSING

Dry powder processing is one of the prominent techniques as it provides control on microstructure and shape forming competency, which is used to process FGMs. This technique involves three basic steps, which led to the development of FGM. Initially, the powder is being weighed and mixed as per the predesigned spatial distribution followed by its functional requirement. After this, the premixed powder is being rammed, stacked, and finally sintered as shown in Figure 3.4. Generally, stepwise structures are formed in this processing method. This technique provides to achieve optimal conditions for graded material having variations in microstructure and compositions. This is done by altering the average particle size of the powder used (Chenglin et al., 1999; Jin, Takeuchi, Honda, Nishikawa, & Awaji, 2005; Zhu et al., 1996).

In this technique, the fabrication of FGM involved the process of graded sintering under ultra-high pressure to fabricate tungsten/copper FGM having 1% lanthanum oxide and 1% titanium carbide, and the specimens demonstrate low porosity. Tungsten/copper FGM exhibits reasonable shear strength between layers and worthy thermal-shock resistance. The FGMs with 1% lanthanum oxide have not deformed, whereas 1% titanium carbide containing FGMs commenced to crack when the powder density of laser is 200 MW-m^2 and tested for thermal-shock resistance (Rajan & Pai, 2014).

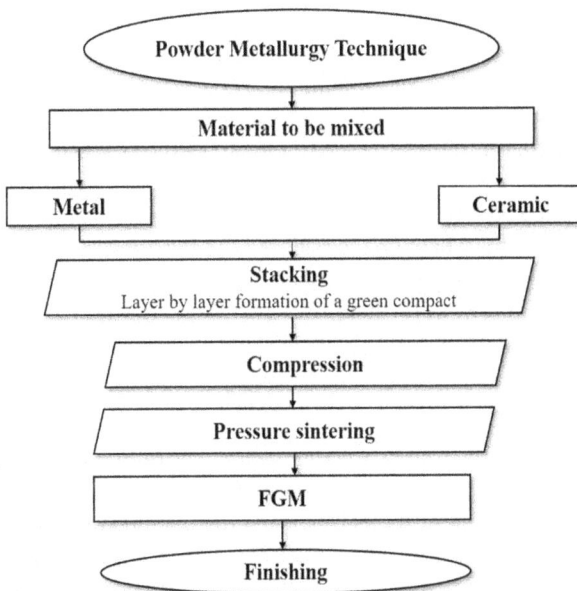

FIGURE 3.4 Fabrication of FGM using powder metallurgy.

3.3.2   SLIP CASTING

The slip casting process involves the filtration where the powder suspension is dispensed into a porous plaster mold. It is one of the conventional powder-based methods which is used in the ceramic industry. Capillary forces are responsible for the removal of liquid from suspension, and the powder particles are enforced to move towards the wall. While carrying out the slip casting procedure, a steep grade will be made by altering the composition or grain dimensions of the applied powder suspension. This technique also necessitates a subsequent consolidation step, where the powder is sintered and a graded structure of the FGM results (Kieback, Neubrand, & Riedel, 2003; Sobczak & Drenchev, 2013).

For example, submicron-sized Al_2O_3 (particle size 0.5 µm) powder and monoclinic zirconium dioxide M-ZrO2 (particle size 0.5 µm) have been selected, and after the sequential casting of slurries, preforms were obtained. Initially, their density was 60% theoretically, but after sintering at 1625°C for 2 h, the density of the sample was 99%. Around 50% of the tetragonal modification of ZrO2 was retained in these preforms, which has been achieved by using the slip casting to assemble preforms followed by sintering. This made it possible to fabricate multilayered composite materials with distinct and well-resolved (≈1 µm) interfaces (Katayama, Sukenaga, Saito, Kagata, & Nakashima, 2011; Kieback et al., 2003; Naebe & Shirvanimoghaddam, 2016).

3.3.3   TAPE CASTING

Tape casting process involves the mixing of solvents and additives to form a free-flowing casting slurry. Thereafter, it is followed by a forming process which is carried out in a tape casting plant where the ceramic slurry is spread continuously over a conveyor belt using a doctor blade to form a thin film. In the end, various other operations such as drying, sintering, and debinding are being carried out for the formation of the final product (Kieback et al., 2003; Saiyathibrahim, Nazirudeen, & Dhanapal, 2015).

After debinding, the green samples were burned at ambient temperature. The sporosity of the multilayer samples is considerably reduced when a low pressure is applied during the self-propagating high-temperature synthesis (SHS) stage and when the green layer thickness is larger than 500 mm. The maintenance of the multilayer structure is significantly reliant on the control of the volume of liquid formed during the SHS reaction. The process setup is shown in Figure 3.5. (Besisa & Ewais, 2016; El-Wazery & El-Desouky, 2015; Jung, Ha, Shin, Hur, & Paik, 2002; Mahamood & Akinlabi, 2017; Yeo, Jung, & Choi, 1998a, 1998b)

3.3.4   CENTRIFUGAL CASTING

It is the most prominently used processing techniques of FGMs. Generally, the centrifugal casting provides better properties depending on the system and solidification conditions. The fabrication of the FGMs using this method depends upon the melting temperature of the reinforcement particle. If the melting point is considerably more compared to the processing temperature, the reinforcement particle retains the solid in a liquid matrix. The only drawback of this processing technique is that it is only apposite for axial symmetry components giving graded structure in the

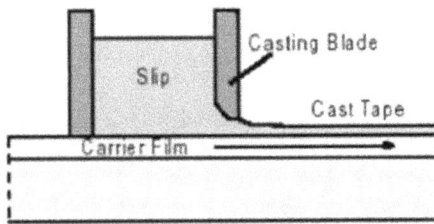

FIGURE 3.5 Tape casting process setup (Saiyathibrahim et al., 2015).

radial direction (El-Hadad, Sato, Miura-Fujiwara, & Watanabe, 2010; Melgarejo, Suárez, & Sridharan, 2008; Watanabe, Shibuya, & Sato, 2013).

In this technique, when the slurry particle is exposed to centrifugal force, it led to the formation of two distinct zones of particles enriched as well as depleted. The melting temperature, densities of the particle and liquid, particle size, and magnitude of centrifugal acceleration direct the extent of particle segregation and relative locations of enriched and depleted particle zones within the casting (Chumanov, Anikeev, & Chumanov, 2015; Melgarejo et al., 2008; Rajan & Pai, 2014; Watanabe & Sato, 2011). The location of heavier as well as lighter particles is decided by their density as the lighter particles move away from the axis of rotation whereas the heavier particles move towards the axis of rotation.

3.3.5 GEL CASTING

Gel casting is one of the best available processing techniques which is used in the formation of compound-shaped dense as well as porous ceramic materials. Various attributes such as short forming time, high green strength, outstanding green machinability, and low-cost machining make it one of the best techniques. The processing technique of gel casting is closely related to shaping methods based on colloidal processes, which was first presented by researchers from O.R. National Laboratory in 1990 (Naebe & Shirvanimoghaddam, 2016; Rajan & Pai, 2014; Saiyathibrahim et al., 2015; Shevchenko, Dudnik, Ruban, Zaitseva, & Lopato, 2003; Zygmuntowicz, Wiecinska, & Miazga, 2018).

3.4 APPLICATION OF FGMs

Since the last three decades, FGMs have been explored in diversified applications such as nuclear reactors, and biomedical and civil industries as shown in Figure 3.6. Their applications are kept on increasing on the grounds that they can be tailored as per the working conditions and requirements. Following are some of the applications in which FGMs have been extensively used.

3.4.1 AEROSPACE

In general, FGMs made of ceramic/metal as outer/inner surface circumvent the unexpected transition between various mechanical properties such as coefficients of thermal expansion and Young's modulus. (Gupta & Talha, 2015; Jha et al., 2013). This

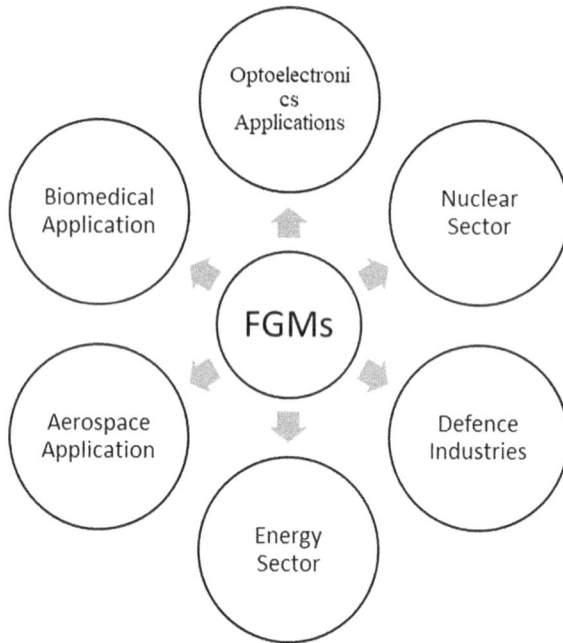

FIGURE 3.6 Applications of functionally graded materials (Gupta & Talha, 2015).

smooth variation of properties not only results in the reduction of thermal stresses and stress concentrations, but also makes it able to withstand very high thermal gradient. These attributes of FGMs offer the best use of such materials in aerospace industries. For instance, when space shuttle re-enters into the Earth's atmosphere, it is exposed to a high-temperature gradient. The outer surface of the space shuttle made of ceramic tiles for thermal protection. These tiles are laminated to the shuttle's superstructure and are susceptible to detachment at the interface. Such problems associated with conventional laminated tiles can be eliminated by using FGMs (Besisa & Ewais, 2016; Mahamood & Akinlabi, 2017).

3.4.2 BIOMEDICAL APPLICATIONS

FGMs have already proven its worth in the field of biomedical. They are very promising materials in biomedical implants such as artificial bone, tooth, skin, knee joints, and hip joints. Living tissues like teeth and bones are considered as natural FGMs. FGMs have established a wide range of applications in dental and orthopedic applications for the replacement of teeth and bones (Ahmed, Rahman, & Adhikary, 2013; Lin et al., 2010; Mehrali et al., 2013; Niu, Rahbar, Farias, & Soboyejo, 2009; Pompe et al., 2003; Watari et al., 2004; Watari, Yokoyama, Saso, & Uo, 1997).

3.4.3 MECHANICAL AND AUTOMOBILE INDUSTRIES

Since last decades, FGMs are widely explored in the manufacturing tools and dyes to have enhanced thermal and wear resistance, reduced scrap, and improved process output. They are also found to be a better replacement of conventional materials for various automobile engine components, flywheels, etc. (Bharti et al., 2013).

3.4.4 DEFENSE

One of the most imperative features of FGMs is the capability to constrain crack propagation. This attribute makes it valuable in a defense application, as penetration resilient materials used for armor plates and bullet-proof vests (Lu, Chekroun, & Abraham, 2011). Other areas of application are tribology, sensors, shape memory alloys (SMAs), etc. (Kawasaki & Watanabe, 2002; Malinina, Sammi, & Gasik, 2005).

Apart from these, numerous applications have been explored since the last three decades and lot more applications are springing up as the processing techniques, mass production, processing cost, and properties of FMG improve.

3.5   GRADATION LAWS OF FGMs

FGMs are advanced composite heterogeneous materials having continuous variations of properties along a desired direction. This continuous gradation has to be governed by certain laws. In the open literature, three laws, i.e., power law, exponential law, and Sigmoid law, have been used extensively to demonstrate the change of properties from one surface to another surface of FGM structures. The mathematical description of these laws is given as follows.

3.5.1   POWER LAW

In this law, the effective material properties $P_{\text{eff}}(z)$ in a preferred direction (generally along the thickness direction) can be given as (Gupta & Talha, 2015; Jha et al., 2013)

$$P_{\text{eff}}(z) = (P_t - P_b)V_f(z) + P_b \tag{3.1}$$

where P_t and P_b are material properties at the top surface and the bottom surface of a FGM structure. The volume fraction $V_f(z)$ is given as

$$V_f = \left(\frac{2z + t}{2t} \right)^p \quad 0 \le p < \infty \tag{3.2}$$

where p is the volume fraction exponent, and t is the thickness of FGM structures.

3.5.2   SIGMOID LAW

This law can be assumed as the modified power law because in this law, the volume fraction of constituent materials is defined using two power-law functions. According to this law, various properties are defined by (Chi & Chung, 2002).

$$\{V_{fr}^1\} = \left\{1 - \frac{1}{2}\left(\frac{(t/2) - z}{(t/2)}\right)^p\right\}$$

$$\{V_{fr}^2\} = \left\{\frac{1}{2}\left(\frac{(t/2) + z}{(t/2)}\right)^p\right\}$$

(3.3)

Using the rule of mixture, the effective material properties of FGMs can be computed as follows:

$$P_{eff}(z) = P_t V_{fr}^1(z) + \left(1 - V_{fr}^1\right)P_b \qquad \text{for } 0 \le z \le t/2$$

$$P_{eff}(z) = P_t V_{fr}^2(z) + \left(1 - V_{fr}^2\right)P_b \qquad \text{for } -t/2 \le z < 0$$

(3.4)

3.5.3 EXPONENTIAL LAW

The effective properties of an exponentially FGM plate are expressed as (Chakraverty & Pradhan, 2014; Fekrar, Houari, Tounsi, & Mahmoud, 2014; Zenkour, Allam, Radwan, & El-Mekawy, 2015)

$$P_{eff} = P_t e^{-\delta\left(1 - \frac{2z}{t}\right)}, \text{ where } \delta = \frac{1}{2}\ln\left(\frac{P_t}{P_b}\right)$$

(3.5)

3.6 HOMOGENIZATION TECHNIQUES OF FGMs

The effective properties of FGMs like advanced composite materials can be obtained using the homogenization technique. Several homogenization techniques are reported in the literature, a few of them are discussed in the following section (Hill, 1965; Klusemann, Böhm, & Svendsen, 2012; Mori & Tanaka, 1973; Shen & Wang, 2012).

3.6.1 THE VOIGT MODEL

This model is very popular while dealing with the structural analysis of FGM. The effective properties (P_{eff}), such as Young's modulus (E_{eff}), Poisson' ratio (v_{eff}), thermal expansion coefficient (α_{eff}), and thermal conductivity (K_{eff}), in a specific direction are determined from the following relations (Gupta, Talha, & Seemann, 2017; Liew, Yang, & Kitipornchai, 2003; Yang & Shen, 2002):

$$E_{eff}(z,T) = \left[E_c(T) - E_m(T)\right]\left(\frac{2z+t}{2t}\right)^p + E_m(T),$$

$$\alpha_{\text{eff}}(z,T) = \left[\alpha_c(T) - \alpha_m(T)\right]\left(\frac{2z+t}{2t}\right)^p + \alpha_m(T),$$

$$\upsilon_{\text{eff}}(z,T) = \left[\upsilon_c(T) - \upsilon_m(T)\right]\left(\frac{2z+t}{2t}\right)^p + \upsilon_m(T)$$

$$\rho_{\text{eff}}(z) = \left[\rho_c - \rho_m\right]\left(\frac{2z+t}{2t}\right)^p + \rho_m,$$

$$K_{\text{eff}}(z,T) = \left[K_c - K_m\right]\left(\frac{2z+t}{2t}\right)^p + K_m$$

(3.6)

3.6.2 THE MORI–TANAKA (M-T) MODEL

According to the M-T scheme (Benveniste, 1987), the E_{eff} and υ_{eff} can be expressed by the following relations:

$$\frac{K_{\text{eff}}(z) - K_t}{K_b - K_t} = \frac{V_b}{1 + \left[1 - V_b\right]\dfrac{3\left[K_b + K_t\right]}{\left[3K_t + 4G_t\right]}}$$

$$\frac{G_{\text{eff}}(z) - G_t}{G_b - G_t} = \frac{G_{\text{eff}}(z)}{1 + \left[1 - V_b\right]\left(\dfrac{G_b + G_t}{G_t + f_1}\right)}$$

(3.7)

Where, $f_1 = \dfrac{G_t\left[9K_t + 8G_t\right]}{\left[6K_t + 2G_t\right]}$

From Equation 3.7, the effective material properties are given as

$$E_{\text{eff}} = \frac{9K_{\text{eff}}G_{\text{eff}}}{3K_f + G_f}, \quad \upsilon_{\text{eff}} = \frac{3K_{\text{eff}} - 2G_{\text{eff}}}{6K_{\text{eff}} + 2G_{\text{eff}}}$$

(3.8)

3.7 MODELS BASED ON PLATE/SHELL THEORIES FOR STRUCTURAL ANALYSIS

The analysis of advanced composite structures is one of the most demanding research domains since last several decades. Authentic and unambiguous estimate of structural response is very necessary to design various structural members for different applications. Abundant plate/shell theories have been established so far to demonstrate structural responses of composite structures. In this viewpoint, Kirchhoff (1850) developed classical plate theory (CPT) in which shear deformation has been neglected. Due to this negligence, its applicability is restricted to the thin plates (Love, 1888; Yang & Shen, 2001). To obliterate this deprivation, the first-order shear deformation theory (FSDT) has been proposed by Reissner (1975) and Mindlin (1951). The FSDT does not incorporate the traction-free boundary conditions; therefore, it necessitates the use of shear correction factor (SCF). The value of SCF depends on several parameters like loading and geometric conditions. To evade the restrictions of CPT and FSDT, various higher-order deformation theories (HSDTs) have been developed. These theories can be classified as polynomial-based HSDT (P-HSDT) and non-polynomial based HSDT (NP-HSDT). In P-HSDTs, Taylor's series expansion is used to govern the shear deformation, whereas, in NP-HSDT, shear-strain function governs the shear deformation.

In this perspective, Levinson (1980), Matsunaga (2008), Pandya and Kant (1988), Reddy (1984), Talha and Singh (2010), Murthy (1981), and Xiang, Jin, Bi, Jiang, and Yang (2011) developed the renowned P-HSDTs which was implemented to analyze the isotropic and advanced composite plates. These theories are predominately used for the structural analysis of composite structures but are difficult to formulate and computationally very expensive.

To assassinate the previously mentioned complexity, NP-HSDTs have been developed which consist of shear-strain functions to demonstrate the realistic deformation. In this regard, Touratier (1991), Aydogdu (2009) Karama, Afaq, and Mistou (2009), Grover, Singh, and Maiti (2013), Adhikari and Singh (2017), Akavci and Tanrikulu (2015), Gupta and Talha (2017), Li, Yu, Han, Zhao, and Wu (2015), Mantari, Bonilla, and Guedes Soares (2014), Mantari and Guedes Soares (2013), Sarangan and Singh (2013), Thai, Ferreira, Abdel Wahab, and Nguyen-Xuan (2016), Tounsi, Houari, and Bessaim (2016), and Thai et al. (2016) are some of the renowned NP-HSDTs in which some trigonometric, logarithmic, and/or exponential functions have been used as -strain function. These theories deliver result close to the exact solution and are less expensive compared to P-HSDTs. Some of the -strain functions available in the literature are given in Table 3.1.

3.8   SOLUTION METHODOLOGIES FOR STRUCTURAL ANALYSIS OF FGMs

The characteristic equations for the structural response (flexural, vibration, and buckling) of beam/plate/shell are generally in the form of single or a system of partial differential equations (PDEs). The analytical solutions (3D exact solution) and approximate analytical methods (power series method, Ritz's method, Galerkin's method, and perturbation method) for such equations are possible for some specific cases (Hassan & Kurgan, 2019). For example, to explore the vibration and buckling response of plates, closed-form solutions (Navier's solution, Levy's solution, etc.) can be used, but its applicability is restricted up to the certain boundary, loading, and environmental conditions. Several literatures are available in which such methods have been used for the analysis (Abualnour, Houari, Tounsi, Adda Bedia, & Mahmoud, 2018; Bodaghi & Saidi, 2010; El-Meiche, Tounsi, Ziane, Mechab, & El, 2011; Gorman, 1979; Mohammadi, Saidi, & Jomehzadeh, 2010; Swaminathan & Patil, 2008; Thai & Kim, 2010). If the structure is of arbitrary shape, then such problems cannot be dealt with the analytical solution. In Figure 3.7, various methods of mathematical modeling and solution methodology have been shown.

TABLE 3.1
Various Shear-Strain Shape Functions

Theory	Shear-Strain Shape Function $f(z)$	Transverse Displacement W
Reissner (1975)	$\dfrac{5z}{h}\left(1-\dfrac{4z^2}{3h^2}\right)$	w_0
Reddy (1984)	$z\left(1-\dfrac{4z^2}{3h^2}\right)$	w_0
Touratier (1991)	$\left(\dfrac{h}{\pi}\right)\sin\left(\dfrac{\pi z}{h}\right)$	w_0
Karama, Afaq, and Mistou (2003)	$ze^{-2\left(z^2/h^2\right)}$	w_0
Ferreira, Roque, and Jorge (2005)	$\mathrm{Sin}\left(\pi z/h\right)$	w_0
Atmane, Tounsi, Mechab, and Bedia (2010)	$\dfrac{\cosh(\pi/2)}{\cosh(\pi/2)-1}z-\dfrac{(h/\pi)\sinh(\pi z/h)}{\cosh(\pi/2)-1}$	w_0
Mantari, Oktem, and Guedes Soares (2012b)	$\tan mz - m\sec^2\left(mz/h\right)z$	w_0
Aydogdu (2009)	$z\alpha^{-2\left(z^2/h^2\right)/\ln\alpha}$	w_0
Mantari, Oktem, and Guedes Soares (2012a)	$\sin\left(\pi z/h\right)e^{m\cos\left(\pi z/h\right)}+\left(m\pi/h\right)z$	w_0
Belabed, Ahmed Mahmoud, and Anwar Beg (2014)	$\sinh(z/h)e^{(1/5h)\cosh(z/h)}+z\left(\dfrac{\cosh(1/2)+(1/5h)\sinh^2(1/2)}{h}e^{(1/5h)\cosh(z/h)}\right)$	$w_b+w_s+\left(1-f'(z)\right)\phi_z$

(Continued)

TABLE 3.1 (Continued)
Various Shear-Strain Shape Functions

Theory	Shear-Strain Shape Function $f(z)$	Transverse Displacement W
Soldatos (1992)	$h\sinh\left(\dfrac{z}{h}\right) - z\cosh\left(\dfrac{1}{2}\right)$	w_0
Lee, Han, and Park (2015)	$\dfrac{4z^2}{3h^2}$	$w_b + w_s + \left(1 - \dfrac{8z}{3h^2}\right)w_z$
Akavci & Tanrikulu, (2015)	$3.7z\left(1.27\,\mathrm{sech}^{0.65}\,(z/h) - 1\right)$	$w_0 + f'(z)\theta_z$
Mantari and Guedes Soares (2014)	$\sinh(z/h)\,e^{m\cosh(z/h)} - z\,\dfrac{\cosh(1/2) + m\sinh^2(1/2)}{h}\,e^{m\cosh(1/2)}$	$w_b + w_s + \cos(z/h)\,\theta$
Neves et al. (2012)	$\sinh(\pi z/h)$	$w_0 + zw_1 + z^2 w_2$
Zenkour (2013)	$h\sinh(z/h) - \left(\dfrac{4z^3}{3h^2}\right)\cosh(1/2)$	$w_0 + (1/12)f'(z)\theta_z$
Grover, Maiti, and Singh (2013)	$\sinh^{-1}\left(rz/h\right) - 2r/h\sqrt{r^2 + 4}$	w
Thai and Choi (2014)	$-\dfrac{z}{4} + \dfrac{5z^3}{3h^2}$	$w_b + w_s + \left(1 - f'(z)\right)\theta_z$
Thai and Choi (2013)	$z - ze^{-2(z/h)^2}$	$w_b + w_s$
Gupta and Talha (2017)	$-\dfrac{h\cosh^2(\kappa/2)}{\sqrt{(1+\kappa^2/4)} - 1}\left[\sinh^{-1}\left(\dfrac{\kappa z}{h}\right) - \left(\dfrac{\kappa}{h}\right)z\right]$	$w_0(x,y,t) + \kappa\cosh^2\left(\dfrac{\kappa z}{h}\right)w_s(x,y,t)$

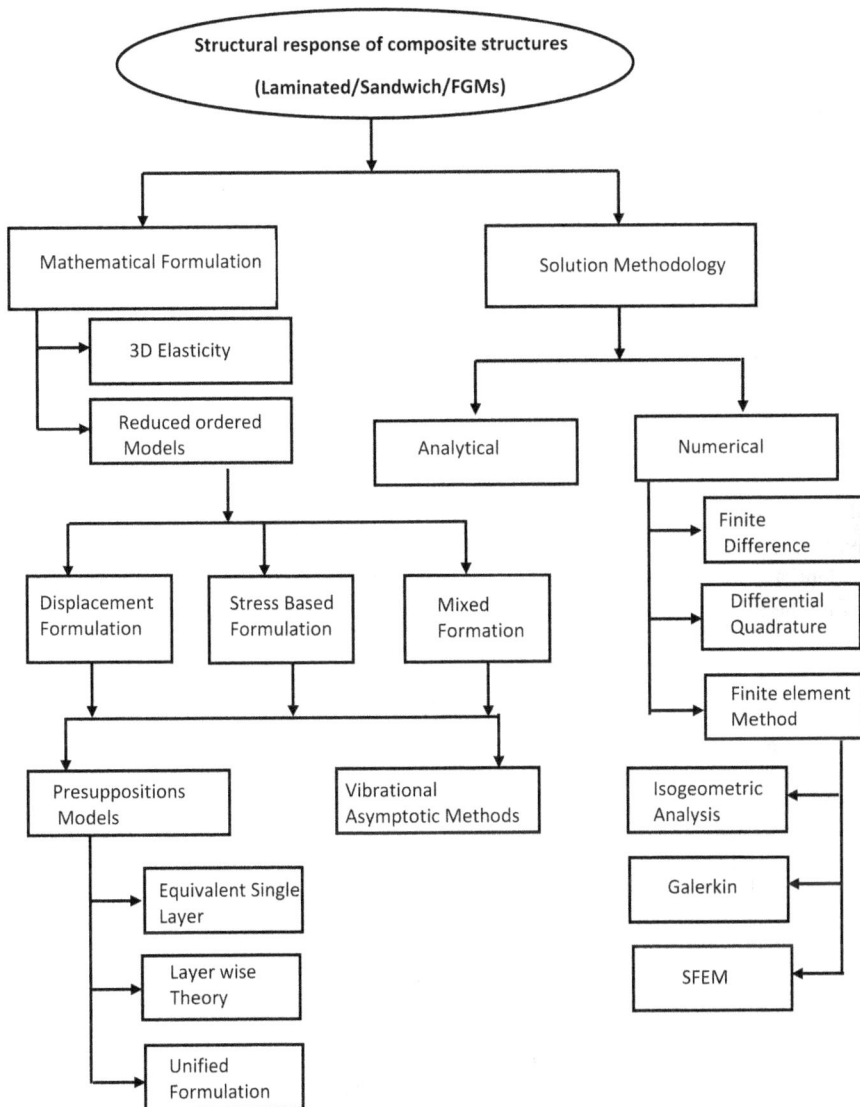

FIGURE 3.7 Various mathematical formulations and solution methodologies for the structural analysis if composite structures (Kumar, 2018).

Therefore, to deal with any type of complex structures, and loading, boundary, and environmental conditions, numerical solutions are predominately adopted. These numerical solutions can be finite element method (FEM), finite volume method (FVM), differential quadrature method (DQM), meshless method, finite strip method (FSM), and finite difference method (FDM). It is noteworthy that these numerical methods have their own specified formulation. Few methods can be combined to achieve the best out of both the methods (Kumar, 2018). For example, the

FEM is implemented for spatial discretization, and the FDM is used for temporal discretization in the dynamic study of composite structures. In the literature, numerous studies have been carried out in which finite element method has been used to find the structural response of structures. Studies carried out by Grover, Maiti, and Singh (2014), Na and Kim (2006), Natarajan, Chakraborty, Ganapathi, and Subramanian (2014), Payette and Reddy, (2014), Pijaudier-Cabot, Bodé, and Huerta (1995), Soh and Wanji (2004), Tran, Thai, and Nguyen-Xuan (2013), and Zafarmand and Kadkhodayan (2015) are the few names in this framework.

3.9 CONCLUSION

A comprehensive review of FGMs has been presented in this chapter in which various types of fabrication techniques, gradation and homogenization techniques, specific applications, and various solution methodologies have been discussed in detail. Some basic models (plate theories) which are generally used for the structural analysis of composite plates have been discussed exclusively. The remarkable potential of FGM lies in the field of aerospace, as well as in nuclear, biomedical, and other applications. FGMs seem to be very promising in the medical field applications such as dentistry and biomedical implants (artificial bones). Still, there is a requirement to refine the fabrication process so that the bulk production of FGMs can be possible.

REFERENCES

Abualnour, M., Houari, M. S. A., Tounsi, A., Adda Bedia, E. A., & Mahmoud, S. R. (2018). A novel quasi-3D trigonometric plate theory for free vibration analysis of advanced composite plates. *Composite Structures*, *184*, 688–697.

Adhikari, B., & Singh, B. (2017). An efficient higher order non-polynomial Quasi 3-D theory for dynamic responses of laminated composite plates. *Composite Structures*. doi: 10.1016/j.compstruct.2017.10.044.

Ahmed, B. T., Rahman, M. Z., & Adhikary, D. (2013). Analysis of Al2O3/Al FGM as biomaterial of artificial human femoral bone and compare with Ti6Al4V alloy through computational study. *Global Journal of Research in Engineering*, *13*(3), 1–8.

Akavci, S. S., & Tanrikulu, A. H. (2015). Static and free vibration analysis of functionally graded plates based on a new quasi-3D and 2D shear deformation theories. *Composites Part B: Engineering*, *83*, 203–215. doi: 10.1016/j.compositesb.2015.08.043.

Atmane, H. A., Tounsi, A., Mechab, I., & Bedia, E. A. A. (2010). Free vibration analysis of functionally graded plates resting on Winkler-Pasternak elastic foundations using a new shear deformation theory. *International Journal of Mechanics and Materials in Design*, *6*(2), 113–121. doi: 10.1007/s10999-010-9110-x.

Aydogdu, M. (2009). A new shear deformation theory for laminated composite plates. *Composite Structures*, *89*(1), 94–101. doi: 10.1016/j.compstruct.2008.07.008.

Belabed, Z., Ahmed Houari, M. S., Tounsi, A., Mahmoud, S. R., & Anwar Beg, O. (2014). An efficient and simple higher order shear and normal deformation theory for functionally graded material (FGM) plates. *Composites Part B: Engineering*, *60*, 274–283. doi: 10.1016/j.compositesb.2013.12.057.

Benveniste, Y. (1987). A new approach to the application of Mori-Tanaka's theory in composite materials. *Mechanics of Materials*, *6*(2), 147–157. doi: 10.1016/0167-6636(87)90005-6.

Besisa, D. H. A., & Ewais, E. M. M. (2016). Advances in functionally graded ceramics – processing, sintering properties and applications. *Advances in Functionally Graded Materials and Structures*, 1–20. doi: 10.5772/62612.

Bever, M. B., & Duwez, P. E. (1972). Gradients in composite materials. *Materials Science and Engineering, 10*, 1–8.

Bharti, I., Gupta, N., & Gupta, K. M. (2013). Novel applications of functionally graded nano, optoelectronic and thermoelectric materials. *International Journal of Materials, Mechanics and Manufacturing, 1*(3), 221–224. doi: 10.7763/IJMMM.2013.V1.47.

Birman, V., & Byrd, L. W. (2007). Modeling and analysis of functionally graded materials and structures. *Applied Mechanics Reviews, 60*(5), 195. doi: 10.1115/1.2777164.

Bodaghi, M., & Saidi, A. R. (2010). Levy-type solution for buckling analysis of thick functionally graded rectangular plates based on the higher-order shear deformation plate theory. *Applied Mathematical Modelling, 34*(11), 3659–3673. doi: 10.1016/j.apm.2010.03.016.

Chakraverty, S., & Pradhan, K. K. (2014). Free vibration of exponential functionally graded rectangular plates in thermal environment with general boundary conditions. *Aerospace Science and Technology, 36*, 132–156. doi: 10.1016/j.ast.2014.04.005.

Chenglin, C., Jingchuan, Z., Zhongda, Y., & Shidong, W. (1999). Hydroxyapatite–Ti functionally graded biomaterial fabricated by powder metallurgy. *Materials Science and Engineering: A, 271*(1–2), 95–100. doi: 10.1016/S0921-5093(99)00152-5.

Chi, S. H., & Chung, Y. L. (2002). Cracking in sigmoid functionally graded coating. *Journal of Mechanics, 18*, 41–53.

Chumanov, I. V., Anikeev, A. N., & Chumanov, V. I. (2015). Fabrication of functionally graded materials by introducing wolframium carbide dispersed particles during centrifugal casting and examination of FGM's structure. *Procedia Engineering, 129*, 816–820. doi: 10.1016/j.proeng.2015.12.111.

El-Hadad, S., Sato, H., Miura-Fujiwara, E., & Watanabe, Y. (2010). Fabrication of Al-Al3Ti/Ti3Al functionally graded materials under a centrifugal force. *Materials, 3*(9), 4639–4656. doi: 10.3390/ma3094639.

El-Meiche, N., Tounsi, A., Ziane, N., Mechab, I., & El, E. A. (2011). A new hyperbolic shear deformation theory for buckling and vibration of functionally graded sandwich plate. *International Journal of Mechanical Sciences, 53*(4), 237–247. doi: 10.1016/j.ijmecsci.2011.01.004.

El-Wazery, M. S., & El-Desouky, A. R. (2015). A review on functionally graded ceramic-metal materials. *Materials and Environmental Science, 6*(5), 1369–1376.

Fekrar, A., Houari, M. S. A., Tounsi, A., & Mahmoud, S. R. (2014). A new five-unknown refined theory based on neutral surface position for bending analysis of exponential graded plates. *Meccanica, 49*(4), 795–810. doi: 10.1007/s11012-013-9827-3.

Ferreira, A. J. M., Roque, C. M. C., & Jorge, R. M. N. (2005). Analysis of composite plates by trigonometric shear deformation theory and multiquadrics. *Computers & Structures, 83*(27), 2225–2237. doi: 10.1016/j.compstruc.2005.04.002.

Gorman, D. J. (1979). Solutions of the Lévy type for the free vibration analysis of diagonally supported rectangular plates. *Journal of Sound and Vibration, 66*(2), 239–246. doi: 10.1016/0022-460X(79)90669-2.

Grover, N., Maiti, D. K., & Singh, B. N. (2013). A new inverse hyperbolic shear deformation theory for static and buckling analysis of laminated composite and sandwich plates. *Computers & Structures, 95*, 667–675. doi: 10.1016/j.compstruct.2012.08.012.

Grover, N., Maiti, D. K., & Singh, B. N. (2014). An efficient C0 finite element modeling of an inverse hyperbolic shear deformation theory for the flexural and stability analysis of laminated composite and sandwich plates. *Finite Elements in Analysis and Design, 80*, 11–22. doi: 10.1016/j.finel.2013.11.003.

Grover, N., Singh, B. N., & Maiti, D. K. (2013a). Analytical and finite element modeling of laminated composite and sandwich plates: an assessment of a new shear deformation theory for free vibration response. *International Journal of Mechanical Sciences, 67*, 89–99. doi: 10.1016/j.ijmecsci.2012.12.010.

Grover, N., Singh, B. N., & Maiti, D. K. (2013b). New nonpolynomial shear-deformation theories for structural behavior of laminated-composite and sandwich plates. *AIAA Journal*, *51*(8), 1861–1871. doi: 10.2514/1.J052399.

Gupta, A., & Talha, M. (2015). Recent development in modeling and analysis of functionally graded materials and structures. *Progress in Aerospace Sciences*, *79*, 1–14. doi: 10.1016/j.paerosci.2015.07.001.

Gupta, A., & Talha, M. (2017). Influence of porosity on the flexural and vibration response of gradient plate using nonpolynomial higher-order shear and normal deformation theory. *International Journal of Mechanics and Materials in Design*. doi: 10.1007/s10999-017-9369-2.

Gupta, A., Talha, M., & Seemann, W. (2017). Free vibration and flexural response of functionally graded plates resting on Winkler–Pasternak elastic foundations using nonpolynomial higher order shear and normal deformation theory. *Mechanics of Advanced Materials and Structures*, *6494*(April), 523–538. doi: 10.1080/15376494.2017.1285459.

Hassan, A. H. A., & Kurgan, N. (2019). A review on buckling analysis of functionally graded plates under thermo- mechanical loads. *International Journal of Engineering & Applied Science*, *11*(1), 345. doi: 10.24107/ijeas.555719.

He, L. H., & Swain, M. V. (2009). Enamel-A functionally graded natural coating. *Journal of Dentistry*, *37*(8), 596–603. doi: 10.1016/j.jdent.2009.03.019.

Hill, R. (1965). A self-consistent mechanics of composite materials. *Journal of the Mechanics and Physics of Solids*, *13*(4), 213–222. doi: 10.1016/0022-5096(65)90010-4.

Jha, D. K., Kant, T., & Singh, R. K. (2013). A critical review of recent research on functionally graded plates. *Composite Structures*, *96*, 833–849. doi: 10.1016/j.compstruct.2012.09.001.

Jin, G., Takeuchi, M., Honda, S., Nishikawa, T., & Awaji, H. (2005). Properties of multi-layered mullite/Mo functionally graded materials fabricated by powder metallurgy processing. *Materials Chemistry and Physics*, *89*(2–3), 238–243. doi: 10.1016/j.matchemphys.2004.03.031.

Jung, Y. G., Ha, C. G., Shin, J. H., Hur, S. K., & Paik, U. (2002). Fabrication of functionally graded ZrO2/NiCrAlY composites by plasma activated sintering using tape casting and it's thermal barrier property. *Materials Science and Engineering A*, *323*(1–2), 110–118. doi: 10.1016/S0921-5093(01)01371-5.

Karama, M., Afaq, K. S., & Mistou, S. (2003). Mechanical behaviour of laminated composite beam by the new multi-layered laminated composite structures model with transverse shear stress continuity. *International Journal of Solids and Structures*, *40*(6), 1525–1546. doi: 10.1016/S0020-7683(02)00647-9.

Karama, M., Afaq, K. S., & Mistou, S. (2009). A new theory for laminated composite plates. *Proceedings of the Institution of Mechanical Engineers, Part L: Journal of Materials: Design and Applications*, *223*(2), 53–62. doi: 10.1243/14644207JMDA189.

Katayama, T., Sukenaga, S., Saito, N., Kagata, H., & Nakashima, K. (2011). Fabrication of Al2O3-W functionally graded materials by slipcasting method. In *IOP Conference Series: Materials Science and Engineering* (Vol. 18, p. 202023). doi: 10.1088/1757-899X/18/20/202023.

Kawasaki, A., & Watanabe, R. J. E. F. M. (2002). Thermal fracture behavior of metal/ceramic functionally graded materials. *Engineering Fracture Mechanics*, *69*, 1713–1728.

Kieback, B., Neubrand, A., & Riedel, H. (2003). Processing techniques for functionally graded materials. *Materials Science and Engineering A*, *362*(1–2), 81–105. doi: 10.1016/S0921-5093(03)00578-1.

Kirchhoff, G. R. (1850). Uber das gleichgewicht und die bewegung einer elastischen Scheibe. *Journal for Pure and Applied Mathematics*, *40*, 51–88.

Klusemann, B., Böhm, H. J., & Svendsen, B. (2012). Homogenization methods for multi-phase elastic composites with non-elliptical reinforcements: comparisons and benchmarks. *European Journal of Mechanics – A/Solids*, *34*, 21–37. doi: 10.1016/j.euromechsol.2011.12.002.

Koizumi, M. (1997). FGM activities in Japan. *Composites Part B: Engineering, 28*(1–2), 1–4. doi: 10.1016/S1359-8368(96)00016-9.

Koizumi, M., & Niino, M. (1995). Overview of FGM research in Japan. *MRS Bulletin, 20*(1), 19–21. doi: 10.1557/S0883769400048867.

Kumar, S. K. (2018). Review of laminated composite plate theories, with emphasis on variational asymptotic method. *AIAA Journal*, 1–7. doi: 10.2514/1.J057552.

Lee, W. H., Han, S. C., & Park, W. T. (2015). A refined higher order shear and normal deformation theory for E-, P-, and S-FGM plates on Pasternak elastic foundation. *Composite Structures, 122*, 330–342. doi: 10.1016/j.compstruct.2014.11.047.

Levinson, M. (1980). An accurate simple theory of statics and dynamics of elastic plates. *Mechanics Research Communications, 7*(6), 343–350.

Li, X., Yu, K., Han, J., Zhao, R., & Wu, Y. (2015). A piecewise shear deformation theory for free vibration of composite and sandwich panels. *Composite Structures, 124*, 111–119. doi: 10.1016/j.compstruct.2015.01.007.

Liew, K. M., Yang, J., & Kitipornchai, S. (2003). Postbuckling of piezoelectric FGM plates subject to thermo-electro-mechanical loading. *International Journal of Solids and Structures, 40*(15), 3869–3892. doi: 10.1016/S0020-7683(03)00096-9.

Lin, D., Li, Q., Li, W., & Swain, M. (2010). Bone remodeling induced by dental implants of functionally graded materials. *Journal of Biomedical Materials Research – Part B Applied Biomaterials, 92*(2), 430–438. doi: 10.1002/jbm.b.31531.

Love, A. E. H. (1888). The small free vibrations and deformation of a thin elastic shell. *Philosophical Transactions of the Royal Society A: Mathematical, Physical and Engineering Sciences, 179*, 491–546. doi: 10.1098/rsta.1888.0016.

Lu, L., Chekroun, M., Abraham, O., Maupin, V., & Villain, G. (2011). Mechanical properties estimation of functionally graded materials using surface waves recorded with a laser interferometer. *NDT & E International, 44*(2), 169–177.

Mahamood, R. M., & Akinlabi, E. T. (2015). Modelling of process parameters influence on degree of porosity in laser-metal deposition process. *Transactions on Engineering Technologies*, 31–42. doi: 10.1007/978-94-017-9588-3_3

Mahamood, R. M., & Akinlabi, E. T. (2017). Types of functionally graded materials and their areas of application. In *Functionally Graded Materials* (pp. 9–21). Cham: Springer International Publishing. doi: 10.1007/978-3-319-53756-6_2.

Malinina, M., Sammi, T., & Gasik, M. M. (2005). Corrosion resistance of homogeneous and FGM coatings. *Materials Science Forum, 492*, 305–310.

Mantari, J. L., Bonilla, E. M., & Guedes Soares, C. (2014). A new tangential-exponential higher order shear deformation theory for advanced composite plates. *Composites Part B: Engineering, 60*, 319–328. doi: 10.1016/j.compositesb.2013.12.001.

Mantari, J. L., & Guedes Soares, C. (2013). Finite element formulation of a generalized higher order shear deformation theory for advanced composite plates. *Composite Structures, 96*, 545–553. doi: 10.1016/j.compstruct.2012.08.004.

Mantari, J. L., & Guedes Soares, C. (2014). A trigonometric plate theory with 5-unknowns and stretching effect for advanced composite plates. *Composite Structures, 107*, 396–405. doi: 10.1016/j.compstruct.2013.07.046.

Mantari, J. L. L., Oktem, A. S., & Guedes Soares, C. (2012a). Bending response of functionally graded plates by using a new higher order shear deformation theory. *Composite Structures, 94*(2), 714–723. doi: 10.1016/j.compstruct.2011.09.007.

Mantari, J. L., Oktem, A. S., & Guedes Soares, C. (2012b). A new trigonometric shear deformation theory for isotropic, laminated composite and sandwich plates. *International Journal of Solids and Structures, 49*(1), 43–53. doi: 10.1016/j.ijsolstr.2011.09.008.

Matsunaga, H. (2008). Free vibration and stability of functionally graded plates according to a 2-D higher-order deformation theory. *Composite Structures, 82*(4), 499–512. doi: 10.1016/j.compstruct.2007.01.030.

Mehrali, M., Shirazi, F. S., Mehrali, M., Metselaar, H. S. C., Kadri, N. A. B., & Osman, N. A. A. (2013). Dental implants from functionally graded materials. *Journal of Biomedical Materials Research – Part A*, *101*(10), 3046–3057. doi: 10.1002/jbm.a.34588.

Melgarejo, Z. H., Suárez, O. M., & Sridharan, K. (2008). Microstructure and properties of functionally graded Al-Mg-B composites fabricated by centrifugal casting. *Composites Part A: Applied Science and Manufacturing*, *39*(7), 1150–1158. doi: 10.1016/j.compositesa.2008.04.002.

Mindlin, R. D. (1951). Influence of rotatory inertia and shear on flexural motions of isotropic, elastic plates. *ASME Journal of Applied Mechanics*, *18*, 31–38.

Mohammadi, M., Saidi, A. R., & Jomehzadeh, E. (2010). Levy solution for buckling analysis of functionally graded rectangular plates. *Applied Composite Materials*, *17*(2), 81–93. doi: 10.1007/s10443-009-9100-z.

Mojdehi, A. R., & Darvizeh, A. (2011). Three dimensional static and dynamic analysis of thick functionally graded plates by the meshless local Petrov–Galerkin (MLPG) method. *Computational Methods in Civil Engineering*, *2*(1), 65–81. Retrieved from http://www.sciencedirect.com/science/article/pii/S0955799711001093.

Mori, T., & Tanaka, K. (1973). Average stress in matrix and average elastic energy of materials with misfitting inclusions. *Acta Metallurgica*, *21*(5), 571–574. doi: 10.1016/0001-6160(73)90064-3.

Murthy, M. V. V. (1981). An improved transverse shear deformation theory for laminated anisotropic plates. *NASA Technical Paper 1903*, (November).

Na, K.-S., & Kim, J.-H. (2006). Thermal postbuckling investigations of functionally graded plates using 3-D finite element method. *Finite Elements in Analysis and Design*, *42*(8–9), 749–756. doi: 10.1016/j.finel.2005.11.005.

Naebe, M., & Shirvanimoghaddam, K. (2016). Functionally graded materials: a review of fabrication and properties. *Applied Materials Today*, *5*(December), 223–245. doi: 10.1016/j.apmt.2016.10.001.

Natarajan, S., Chakraborty, S., Ganapathi, M., & Subramanian, M. (2014). A parametric study on the buckling of functionally graded material plates with internal discontinuities using the partition of unity method. *European Journal of Mechanics, A/Solids*, *44*, 136–147. doi: 10.1016/j.euromechsol.2013.10.003.

Neves, A. M. A., Ferreira, A. J. M., Carrera, E., Cinefra, M., Roque, C. M. C., Jorge, R. M. N., & Soares, C. M. M. (2012). A quasi-3D hyperbolic shear deformation theory for the static and free vibration analysis of functionally graded plates. *Composite Structures*, *94*(5), 1814–1825. doi: 10.1016/j.compstruct.2011.12.005.

Niu, X., Rahbar, N., Farias, S., & Soboyejo, W. (2009). Bio-inspired design of dental multilayers: Experiments and model. *Journal of the Mechanical Behavior of Biomedical Materials*, *2*(6), 596–602. doi: 10.1016/j.jmbbm.2008.10.009.

Palmer, L. C., Newcomb, C. J., Kaltz, S. R., Spoerke, E. D., & Stupp, S. I. (2008). Biomimetic systems for hydroxyapatite mineralization inspired by bone and enamel. *Chemical Review*, *108*, 4754–4783.

Pandya, B. N., & Kant, T. (1988). Finite element analysis of laminated composite plates using a higher-order displacement model. *Composites Science and Technology*, *32*(2), 137–155. doi: 10.1016/0266-3538(88)90003-6.

Payette, G. S., & Reddy, J. N. (2014). A seven-parameter spectral/hp finite element formulation for isotropic, laminated composite and functionally graded shell structures. *Computer Methods in Applied Mechanics and Engineering*, *278*, 664–704. doi: 10.1016/j.cma.2014.06.021.

Pijaudier-Cabot, G., Bodé, L., & Huerta, A. (1995). Arbitrary Lagrangian-Eulerian finite element analysis of strain localization in transient problems. *International Journal for Numerical Methods in Engineering*, *38*(24), 4171–4191. doi: 10.1002/nme.1620382406.

Pompe, W., Worch, H., Epple, M., Friess, W., Gelinsky, M., Greil, P., … Schulte, K. (2003). Functionally graded materials for biomedical applications. *Materials Science and Engineering A, 362*(1–2), 40–60. doi: 10.1016/S0921-5093(03)00580-X.

Rajan, T. P. D., & Pai, B. C. (2014). Developments in processing of functionally gradient metals and metal – ceramic composites : a review. *Acta Metallurgica Sinica, 27*(5), 825–838. doi: 10.1007/s40195-014-0142-3.

Reddy, J. N. (1984). A simple higher-order theory for laminated composite plates. *Journal of Applied Mechanics, 51*(4), 745. doi: 10.1115/1.3167719.

Reissner, E. (1975). On transverse bending of plates, including the effect of transverse shear deformation. *International Journal of Solids and Structures, 11*(5), 569–573. doi: 10.1016/0020-7683(75)90030-X.

Saiyathibrahim, A., M Nazirudeen, S. S., & Dhanapal, P. (2015). Processing techniques of functionally graded materials – a review. In *Proceedings of the International Conference on Systems, Science, Control, Communication, Engineering and Technology* (Vol. 01, pp. 98–105).

Sarangan, S., & Singh, B. N. (2013). Higher-order closed-form solution for the analysis of laminated composite and sandwich plates based on new shear deformation theories. *Composite Structures, 75*, 324–336. doi: 10.1016/j.compstruct.2015.11.049.

Shen, H.-S., & Wang, Z.-X. (2012). Assessment of voigt and Mori-Tanaka models for vibration analysis of functionally graded plates. *Composite Structures, 94*, 2197–2208. doi: 10.1016/j.compstruct.2012.02.018.

Shevchenko, A. V., Dudnik, E. V, Ruban, A. K., Zaitseva, Z. A., & Lopato, L. M. (2003). Refractory and ceramics materials functional graded materials based on ZrO2 and Al2O3 production methods. *Powder Metallurgy and Metal Ceramics, 42*, 145–153.

Silva, E. C. N., Walters, M. C., Paulino, G. H. (2006). Modeling bamboo as a functionally graded material: lessons for the analysis of affordable materials. *Journal of Materials Science, 41*, 6991–7004.

Sobczak, J. J., & Drenchev, L. (2013). Metallic functionally graded materials: a specific class of advanced composites. *Journal of Materials Science and Technology, 29*(4), 297–316. doi: 10.1016/j.jmst.2013.02.006.

Soh, A.-K., & Wanji, C. (2004). Finite element formulations of strain gradient theory for microstructures and theC0–1 patch test. *International Journal for Numerical Methods in Engineering, 61*(3), 433–454. doi: 10.1002/nme.1075.

Soldatos, K. P. (1992). A transverse shear deformation theory for homogenous monoclinic plates. *Acta Mechanica, 94*, 195–220.

Suk, M. J., Choi, S. I., Kim, J. S., Kim, Y. D., & Kwon, Y. S. (2003). Fabrication of a porous material with a porosity gradient by a pulsed electric-current sintering process. *Metals and Materials International, 9*, 599–603.

Swaminathan, K., & Patil, S. S. (2008). Analytical solutions using a higher order refined computational model with 12 degrees of freedom for the free vibration analysis of antisymmetric angle-ply plates. *Composite Structures, 82*(2), 209–216. doi: 10.1016/j.compstruct.2007.01.001.

Talha, M., & Singh, B. N. (2010). Static response and free vibration analysis of FGM plates using higher order shear deformation theory. *Applied Mathematical Modelling, 34*(12), 3991–4011. doi: 10.1016/j.apm.2010.03.034.

Thai, C. H., Ferreira, A. J. M., Abdel Wahab, M., & Nguyen-Xuan, H. (2016). A generalized layerwise higher-order shear deformation theory for laminated composite and sandwich plates based on isogeometric analysis. *Acta Mechanica, 227*(5), 1225–1250. doi: 10.1007/s00707-015-1547-4.

Thai, H.-T., & Choi, D.-H. (2013). Efficient higher-order shear deformation theories for bending and free vibration analyses of functionally graded plates. *Archive of Applied Mechanics, 83*(12), 1755–1771. doi: 10.1007/s00419-013-0776-z.

Thai, H.-T., & Choi, D.-H. (2014). Improved refined plate theory accounting for effect of thickness stretching in functionally graded plates. *Composites Part B: Engineering, 56,* 705–716. doi: 10.1016/j.compositesb.2013.09.008.

Thai, H.-T., & Kim, S.-E. (2010). Free vibration of laminated composite plates using two variable refined plate theory. *International Journal of Mechanical Sciences, 52*(4), 626–633. doi: 10.1016/j.ijmecsci.2010.01.002.

Tounsi, A., Houari, M. S. A., & Bessaim, A. (2016). A new 3-unknowns non-polynomial plate theory for buckling and vibration of functionally graded sandwich plate. *Structural Engineering and Mechanics, 60*(4), 547–565. doi: 10.12989/sem.2016.60.4.547.

Touratier, M. (1991). An efficient standard plate theory. *International Journal of Engineering Science, 29*(8), 901–916.

Tran, L. V., Thai, C. H., & Nguyen-Xuan, H. (2013). An isogeometric finite element formulation for thermal buckling analysis of functionally graded plates. *Finite Elements in Analysis and Design, 73,* 65–76. doi: 10.1016/j.finel.2013.05.003.

Watanabe, Y., & Sato, H. (2011). Review fabrication of functionally graded materials under a centrifugal force. *Nanocomposites with Unique Properties and Applications in Medicine and Industry,* 133–150. doi: 10.5772/1549.

Watanabe, Y., Shibuya, M., & Sato, H. (2013). Fabrication of Al/diamond particles functionally graded materials by centrifugal sintered-casting method. *Journal of Physics: Conference Series, 419,* 012002. doi: 10.1088/1742-6596/419/1/012002.

Watari, F., Yokoyama, A., Omori, M., Hirai, T., Kondo, H., Uo, M., & Kawasaki, T. (2004). Biocompatibility of materials and development to functionally graded implant for bio-medical application. *Composites Science and Technology, 64*(6), 893–908. doi: 10.1016/j.compscitech.2003.09.005.

Watari, F., Yokoyama, A., Saso, F., & Uo, M. (1997). Fabrication and properties of functionally graded dental implant. *Composites Part B, 28B,* 5–11.

Xiang, S., Jin, Y., Bi, Z., Jiang, S., & Yang, M. (2011). A n-order shear deformation theory for free vibration of functionally graded and composite sandwich plates. *Composite Structures, 93*(11), 2826–2832. doi: 10.1016/j.compstruct.2011.05.022.

Yang, J., & Shen, H. S. (2001). Dynamic response of initially stressed functionally graded rectangular thin plates. *Composite Structures, 54*(4), 497–508. doi: 10.1016/S0263-8223(01)00122-2.

Yang, J., & Shen, H. S. (2002). Vibration characteristics and transient response of shear-deformable functionally graded plates in thermal environments. *Journal of Sound and Vibration, 255,* 579–602. doi: 10.1006/jsvi.2001.4161.

Yeo, J. G., Jung, Y. G., & Choi, S. C. (1998a). Design and microstructure of ZrO$_2$/SUS316 functionally graded materials by tape casting. *Materials Letters, 37*(6), 304–311. doi: 10.1016/S0167-577X(98)00111-6.

Yeo, J. G., Jung, Y. G., & Choi, S. C. (1998b). Zirconia-stainless steel functionally graded material by tape casting. *Journal of the European Ceramic Society, 18,* 1281–1285.

Zafarmand, H., & Kadkhodayan, M. (2015). Free vibration analysis of thick disks with variable thickness containing orthotropic-nonhomogeneous material using finite element method. *Journal of Theoretical and Applied Mechanics, 53*(4), 1005–1018. doi: 10.15632/jtam-pl.53.4.1005.

Zenkour, A. M. (2013). A simple four-unknown refined theory for bending analysis of functionally graded plates. *Applied Mathematical Modelling, 37*(20), 9041–9051. doi: 10.1016/j.apm.2013.04.022.

Zenkour, A. M., Allam, M. N. M., Radwan, A. F., & El-Mekawy, H. F. (2015). Thermo-mechanical bending response of exponentially graded thick plates resting on elastic foundations. *International Journal of Applied Mechanics, 7*(4), 1–24. doi: 10.1142/S1758825115500623.

Zhu, J. C., Yin, Z. D., & Lai, Z. H. (1996). Fabrication and microstructure of ZrO_2-Ni functional gradient material by powder metallurgy. *Journal of Materials Science, 31*(21), 5829–5834. doi: 10.1007/BF01160836.

Zygmuntowicz, J., Wiecinska, P., & Miazga, A. (2018). Thermoanalytical studies of the ceramic-metal composites obtained by gel-centrifugal casting. *Journal of Thermal Analysis and Calorimetry, 133*(1), 303–312. doi: 10.1007/s10973-017-6647-z.

4 Heat Transfer in Noncircular Microchannels Using Water and Ethylene Glycol as Base Fluids with Zn and ZnO Nanoparticles

Monoj Bardalai, Md. Adam Yamin,
Bhaskarjyoti Gogoi, and Rajeev Goswami
Tezpur University

CONTENTS

4.1 BACKGROUND OF PROBLEM

The removal of heat from the microprocessors of high-speed computers is the recent interest among the researchers. Over the last few decades, microchannels have been found to be used for heat removal purpose in electronic devices. The channels with dimension in the range of 1 mm <1 μ are known as microchannels. The surface-to-volume (s/v) ratio in microchannel is quite high. The high s/v ratio enhances heat transfer rate, and thus, the devices with microchannel have become very useful part in compact heat exchangers. Mala and Di [1] carried out the experiment to investigate the flow characteristics of water in microtube. The results of the study show that at higher Re, pressure drop is significantly higher in comparison with the Poiseuille flow theory. Jung and Kwak [2] carried out the experiments on the water flow in rectangular microchannel to see the influence of heat transfer coefficient and friction factor. According to this study, the friction factor in the microchannel is close to the theoretical value, while the heat transfer mechanism depends on fluid properties such as viscosity.

The improvement of heat transfer in microchannel can be done by enhancing fluid and geometry of the channel. Enhanced fluids prepared by dispersing some nanoparticles into a base fluid such as water and ethylene glycol are known as nanofluids. The highly thermal conductive materials such as carbon, metals, and metal oxides are selected as nanoparticles. Therefore, nanofluids possess higher thermal conductivity and heat transfer coefficient as compared to the base fluids. These advantages of nanofluids have motivated the researchers to investigate thermal performance of microchannel using different nanofluids.

Nanofluids are prepared by adding metallic (Al, Zn, Si, etc.) or metal oxide (Al_2O_3, ZnO, SiO_2, etc.) nanoparticles with the conventional water and ethylene glycol in order to enhance the performance of heat transfer. It is observed that optimum concentration and size of nanoparticle result in effective rate of heat transfer. This motivated Tukerman and Pease [3] to develop microchannel heat sink (MCHS) about 20 years back. It was found that heat removal in circular duct is higher than that in noncircular ducts; however, pressure drop is also found to be higher in the circular duct as compared to noncircular duct [4]. Out of these, performance of heat transfer was found to be highest in triangular duct [5].

The pressure drop in nanofluids is found to be smaller as compared to the microfluids [6]. Khaled et al. [7] performed the study of heat transfer on base fluids and nanofluids flow in the laminar regime of a channel. They showed the results of significant enhancement of heat transfer due to the application of nanoparticles with the base fluid. Lotfi et al. [8] used different numerical methods such as Eulerian, one-phase-, and two-phase-mix methods to explore the impact of large range of nanoparticle volume fractions on the parameters related to heat transfer. Their study revealed that the rate of heat transfer increases by increasing particle concentration. Ahmad et al. [9] performed the numerical study on heat transfer with nanoparticles such as alumina, copper, and silicon oxide nanoparticles in ethylene glycol using triangular channel. The study showed about 50% increase in Nu when Re was varied from 100 to 800. The extensive study of the literature shows that pressure drop and rate of heat transfer are less in noncircular channel as compared to the circular channel. He et al. [10] performed the study on rectangular, triangular, and trapezoidal microchannels. They found that heat transfer rate was more in rectangular microchannel in comparison with the other types of microchannels.

Azizi et al. [11] studied the thermal behavior of Cu-water (Cu/W) nanofluid flow through a rectangular microchannel coupled with a cylindrical shape. Their study revealed that Nu can be enhanced up to 23% by the addition of 0.3 wt.% nanoparticles. It was found that applying 4.5% concentrated CuO-water (CuO/W) nanofluid in a rectangular microchannel increases the pressure drop by 70% [12]. A comparative study of nine different nanofluids with three different base fluids was carried out by Kalteh et al. [13]. The diamond–water nanofluid shows the highest heat transfer coefficient, whereas the lowest heat transfer coefficient is found in SiO_2-water nanofluid.

The performance of thermal behavior of nanofluids is characterized by thermal diffusion, Prandtl number (Pr), and Re, which are the functions of properties of the nanofluids. Therefore, the study of nanofluid properties has become very important for investigating thermal performance of nanofluids. A correlation was suggested by Vajjha and Das [14] in order to determine the specific heat of fluids containing nanoparticles. They used the nanoparticles such as Al_2O_3, ZnO, and SiO_2 with a base fluid which was the mixture of ethylene glycol and water in the ratio of 60:40 by mass basis. Satti et al. [15] also evaluated the specific heat of propylene glycol–water nanofluids by dispersing different metal oxide nanoparticles. The correlations proposed by Esfe and Saedodin [16,17] can also be used for the evaluation of viscosity and thermal conductivity of nanofluid such as ZnO-EG. Yu et al. [18] observed Newtonian behavior in lower volume fraction of ZnO-EG nanofluid, whereas shear-thinning behavior was found at higher volume fraction. In the research work carried out by Prajapati et al. [19], the heat transfer coefficient was found to be increased by increasing the volume fraction of ZnO nanoparticles in water. The flow of ethylene glycol (EG) with different nanoparticles through a microtube was investigated by Salman et al. [20]. According to their study, the base fluid EG with SiO_2 nanoparticles possesses the highest Nu among ZnO, CuO, Al_2O_3, and pure EG.

Hosseini et al. [21] investigated the influence of heat transfer by magnetic field in microchannel heat sink. They used Al_2O_3/W as nanofluid for the study and discussed the influence of volume fraction and size of nanoparticles, porosity, aspect ratio of the channel, and Hartmann number on velocity, temperature distribution, and Nu. Another study on heat transfer analysis in microchannel heat sink using nanofluid embedded with pin fins was performed by Zargartalebi and Azaiez in 2018 [22]. The study shows that the distribution of nanoparticles, the geometry of the microchannel, and properties of nanoparticles play important roles on heat transfer. Recently, Kahani [23] performed the heat transfer analysis of Al_2O_3/W nanofluid in rectangular microchannel heat sink at constant heat flux boundary condition. Modified dispersion model was used in their study to examine the heat transfer behavior of nanofluids for different nanoparticles and found the possibility of application of the model in predicting thermal performance nanofluid flow.

4.2 PROBLEM DEFINITION

In this study, heat transfer of four different nanofluids, ZnO/W, ZnO/EG, Zn/W, and Zn/EG in two noncircular microchannels, is investigated using ANSYS R15.0 version. The volume fraction (ϕ) of the nanofluids is in the range of 1%–4%. The Reynolds number of the nanofluids is considered between 20 and 160. The heat transfer of the nanofluids is examined based on two boundary conditions, namely, constant temperature at the wall and constant heat flux.

4.2.1 DESCRIPTION OF THE GEOMETRIC MODEL

The schematic diagrams of the rectangular and triangular microchannels are presented by the plots of Figure 4.1. The hydraulic diameter and length of each microchannel are taken as 0.667 μm and 5 cm, respectively. The hydraulic diameter of the channel is shown in Equation 4.1.

$$D_h = \frac{4A}{P} \tag{4.1}$$

where A is the cross-sectional area, and P is the perimeter of the duct.

The area of cross section and perimeter of rectangular and triangular channels are expressed in the following equations, respectively:

$$A = HW \tag{4.2}$$

$$P = 2(H+W) \tag{4.3}$$

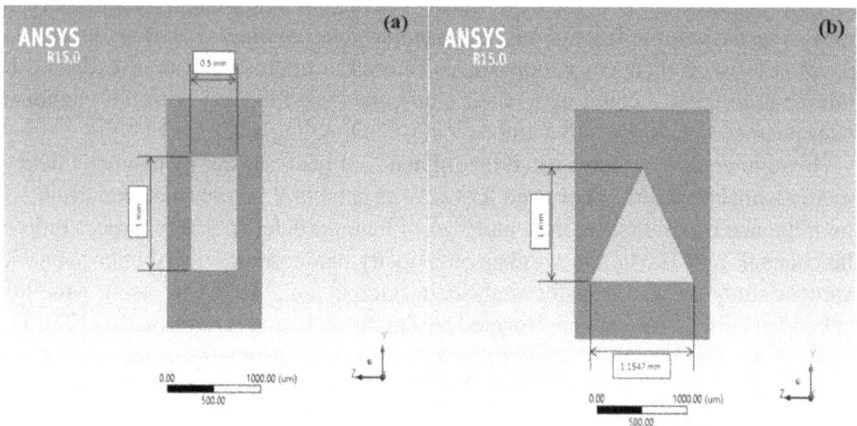

FIGURE 4.1   (a) Rectangular and (b) triangular channels.

$$A = \frac{\sqrt{3}}{4} B^2 \tag{4.4}$$

$$P = 3B \tag{4.5}$$

where H and W are the height and the width of the rectangular channel, respectively, whereas B is the base of the triangular channel.

4.2.2 BOUNDARY CONDITIONS

The temperature of the wall for constant wall boundary condition is considered as 50.15°C (refer Figure 4.2) for both rectangular and triangular microchannels. A constant heat flux 20,000 W/m² is used for water-based nanofluids, whereas 40,000 W/m² is applied when ethylene glycol is a base fluid. For all the conditions, the temperature of the nanofluid at the inlet is maintained at 27°C.

4.2.3 GOVERNING EQUATIONS

The flow of the nanofluids carrying very small nanoparticles (diameter <100 nm) is modeled as single-phase flow [24]. Thus, nanofluid flow can be modeled as the flow of single-phase in this study. Further, the assumptions made for the study are (i) fluid flow and heat transfer are steady and 3D; (ii) the fluid is incompressible and the flow is laminar; (iii) physical properties of the nanofluids do not vary with temperature; (iv) size of the nanoparticles is less than 100 nm; and (v) the effect of buoyancy, viscous dissipation, and heat transfer by radiation are neglected. Based on

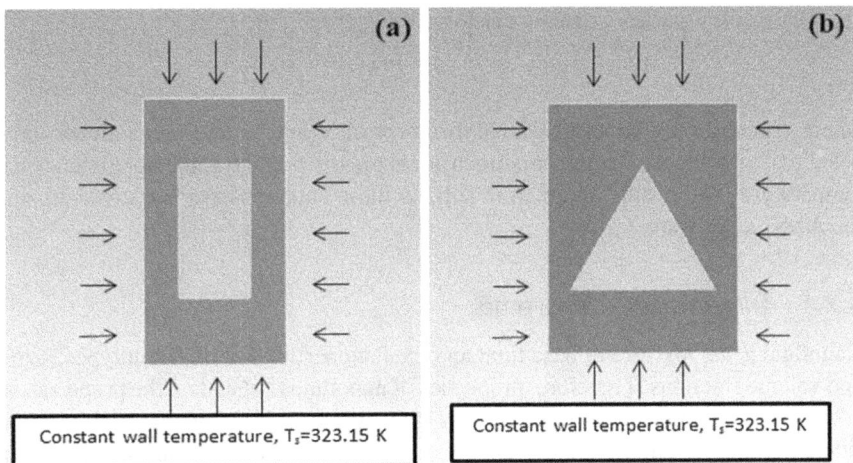

FIGURE 4.2 Constant temperature at the boundary for (a) rectangular channel and (b) triangular channel.

the abovementioned assumptions, the governing equations are given in the following equations:

The equation for mass conservation is

$$\frac{\partial u}{\partial x} + \frac{\partial v}{\partial y} + \frac{\partial w}{\partial z} = 0 \qquad (4.6)$$

The momentum equations for the conservation of momentum in all three directions are presented in the following equations:

The equation of momentum in x-direction is

$$\left(u\frac{\partial u}{\partial x} + v\frac{\partial u}{\partial y} + w\frac{\partial u}{\partial z} \right) = -\frac{dP}{dx} + \frac{1}{Re}\left(\frac{\partial^2 u}{\partial x^2} + \frac{\partial^2 u}{\partial y^2} + \frac{\partial^2 u}{\partial z^2} \right) \qquad (4.7)$$

The equation of momentum in y-direction is

$$\left(u\frac{\partial v}{\partial x} + v\frac{\partial v}{\partial y} + w\frac{\partial w}{\partial z} \right) = -\frac{dP}{dy} + \frac{1}{Re}\left(\frac{\partial^2 v}{\partial x^2} + \frac{\partial^2 v}{\partial y^2} + \frac{\partial^2 v}{\partial z^2} \right) \qquad (4.8)$$

The equation of momentum in z-direction is

$$\left(u\frac{\partial w}{\partial x} + v\frac{\partial w}{\partial y} + w\frac{\partial w}{\partial z} \right) = -\frac{dP}{dz} + \frac{1}{Re}\left(\frac{\partial^2 w}{\partial x^2} + \frac{\partial^2 w}{\partial y^2} + \frac{\partial^2 w}{\partial z^2} \right) \qquad (4.9)$$

The energy equation for the conservation of energy is

$$\left(u\frac{\partial \theta}{\partial x} + v\frac{\partial \theta}{\partial y} + w\frac{\partial \theta}{\partial z} \right) = \frac{1}{Re\,Pr}\left(\frac{\partial^2 \theta}{\partial x^2} + \frac{\partial^2 \theta}{\partial y^2} + \frac{\partial^2 \theta}{\partial z^2} \right) \qquad (4.10)$$

where x, y, and z are nondimensional distances; u, v, and w are the nondimensional velocity components; P is the nondimensional pressure; and θ is the nondimensional temperature. The results obtained in solving these equations are investigated using the ANSYS software.

4.2.4 THERMOPHYSICAL PROPERTIES

Nanofluid is the mixture of base fluid and the nanoparticles of different types, sizes, and volume fractions. Therefore, properties of nanofluids depend on the properties of both base fluid and nanoparticles. In this study, the size of the nanoparticles is considered between 25 and 100 nm. Many researchers have proposed various models for the evaluation of thermophysical properties of nanofluids. Based on these models, the thermophysical properties of nanofluids are calculated using Equations 4.11–4.14 [1,16,24], and the obtained results are shown in Table 4.1.

TABLE 4.1

Thermophysical Properties of Nanofluids

Nanofluids	ϕ (%)	μ_{nf} (Pa·S)	k_{nf} (W/m K)	ρ_{nf} (kg/m³)	$c_{p_{nf}}$ (J/kg·K)
ZnO/W	1	0.001028	0.6175	1044.278	3986.739
	2	0.001055	0.6354	1090.356	3752.073
	4	0.001110	0.6722	1182.512	3492.260
ZnO/EG	1	0.016093	0.2592	2324.334	1156.346
	2	0.016485	0.2670	2240.453	1201.292
	4	0.017387	0.2830	2090.212	1291.184
Zn/W	1	0.001028	0.6179	3925.930	1059.518
	2	0.001053	0.6361	3697.880	1120.836
	4	0.001110	0.6738	3309.250	1243.472
Zn/EG	1	0.016093	0.2590	2290.961	1171.586
	2	1171.586	0.2670	2179.043	1231.772
	4	0.017380	0.2830	1985.090	1352.144

The viscosity of nanofluids depends on the viscosity of base fluid and volume fraction of the nanoparticles. The following equations show the coefficients of viscosity of nanofluids at different volume fractions:

$$\mu_{nf} = (1 + 2.5\phi) \tag{4.11}$$

For $\phi \leq 1\%$,

$$\mu_{nf} = \left\{ \frac{1}{(1-\phi)^{2.5}} \right\} \mu_f \tag{4.12}$$

For $1 < \phi < 4\%$,

The potential of enhancement of the nanofluids is given by thermal conductivity of the nanofluid. The proper expression for evaluating the thermal conductivity is given in the following equation:

$$k_{nf} = \left\{ \frac{k_s + 2k_f + 2\phi(k_s - k_f)}{k_s + 2k_f - \phi(k_s - k_f)} \right\} k_f \tag{4.13}$$

where k_{nf}, k_s, and k_f are the thermal conductivity of the nanofluids, nanoparticles, and base fluid, respectively.

Density of the nanofluid is obtained by knowing the density of both nanoparticles and the base fluid along with nanoparticle concentrations, which is given in the following equation:

$$\rho_{nf} = \rho_s \phi + (1-\phi)\rho_f \tag{4.14}$$

where ρ_{nf} is the density of the nanofluid, whereas ρ_s and ρ_f are the densities of nanoparticles and base fluid, respectively.

Heat capacity of the nanofluid can be determined from the following equation:

$$c_{p_{nf}} = \frac{c_{ps}\phi + (1-\phi)C_{pf}}{\rho_{nf}} \tag{4.15}$$

where $c_{p_{nf}}$, c_{ps}, and c_{pf} are heat capacity of nanofluid, nanoparticles, and base fluids respectively.

Thermophysical properties of the nanofluids were determined using Equations 4.11–4.15 for different volume fractions with the help of ANSYS Workbench Fluent, which are shown in Table 4.1. The mass flow rate ($\dot{m}_{nf}$) of the nanofluid is used in calculating the $\bar{h}$, which is expressed in the following by equation:

$$\dot{m}_{nf} = \rho_{nf} A v \tag{4.16}$$

where v is the velocity of the nanofluids, and A is the area of cross section of the microchannel.

The $\bar{h}$ for both the boundary conditions are calculated using Equations 4.17 and 4.18.

For the condition of constant wall temperature, $\bar{h}$ is given in the following equation:

$$\bar{h} = \frac{\dot{m}_{nf}}{WL} c_{p_{nf}} \ln \frac{\Delta T_i}{\Delta T_0} \tag{4.17}$$

where ΔT_i, ΔT_o, W, and L are inlet temperature difference, outlet temperature difference, width and length of the channels, respectively.

For constant heat flux condition, $\bar{h}$ is calculated by the following equation:

$$\bar{h} = \dot{m}_{nf} \frac{c_{p_{nf}} (T_0 - T_i)}{A_s (T_w - T_b)} \tag{4.18}$$

where A_s and T_w are the heated surface area and wall temperature of the microchannels, respectively, whereas T_i, T_0 and T_b are inlet, outlet, and bulk temperatures of the nanofluids, respectively.

The analysis of heat transfer and fluid flow are performed at different Re, which is determined according to the relation shown in Equation 4.17:

$$Re = \frac{\rho_{nf} v d_h}{\mu_{nf}} \tag{4.19}$$

The Darcy friction factor of any fluid flow is an important parameter indicating the frictional force influenced by pressure loss, density of the nanofluids, velocity, and geometry of the channel. This Darcy friction factor can be evaluated by the following equation:

$$f = 2\frac{D_h \Delta P}{L \rho_{nf} v^2} \tag{4.20}$$

where L and ΔP are length of the microchannel and pressure drop of the nanofluid in the microchannel, respectively.

4.3 RESULTS AND DISCUSSION

The thermal and fluid flow behaviors of the nanofluids, namely, ZnO/W, ZnO/EG, Zn/W, and Zn/EG, through rectangular and triangular microchannels are numerically investigated for both the constant heat flux and constant temperature boundary conditions. In this work, three different nanoparticle volume fractions ($\phi = 1\%$, 2%, and 4%) are used in the nanofluids. The range of Re taken for constant temperature condition was 80–160, whereas for constant heat flux, the range of Re was 20–60.

4.3.1 TEMPERATURE VARIATION

The variation of outlet temperatures of nanofluid in rectangular and triangular ducts at constant wall temperatures is presented in Figure 4.3. It is observed that outlet temperature gradually decreases with the increase in Re, which indicates that heat transfer has increased with Re. The same trend was observed by Kahani [23] for the wall temperature of the microchannel vs. Re by considering both nanofluid and water. In both the microchannels, highest outlet temperatures are found for the highest volume fraction (i.e., $\phi = 4\%$). The decrease in temperature in the case of

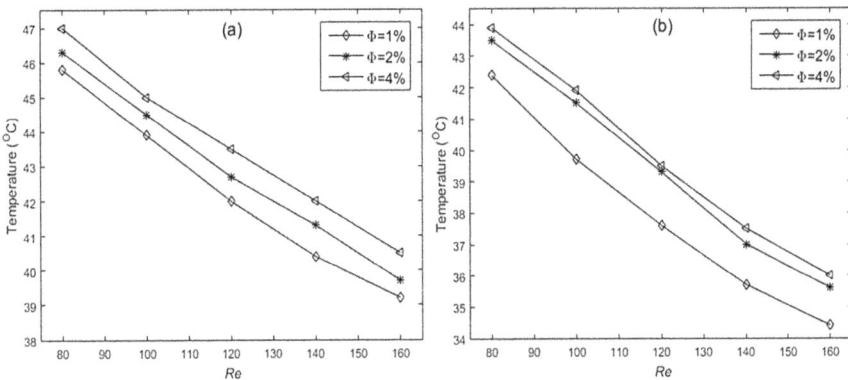

FIGURE 4.3 Variation of outlet temperature with Re in (a) rectangular and (b) triangular ducts.

rectangular duct is found to be about 14%, 15%, whereas in the case of triangular duct, it is near by 18%, 19%. This reveals that increase of heat transfer is more in triangular duct than rectangular duct with the increase in Re.

The temperature difference between wall of the microchannel and bulk nanofluids at constant heat flux for different volume fractions is shown in Figure 4.4a–d. It is seen that in all the nanofluids, the temperature differences (ΔT) decrease with the increase in Re. It is also observed that, in each case, the difference of temperature is highest for the largest volume fraction, whereas it is lowest for the smallest volume fraction. This is an indication of increase of heat transfer with the increase in volume fraction. At low Re, ΔT shows the decreasing trend by increasing Re for all considered nanofluids (refer Figure 4.4). The decrease in ΔT in EG-based nanofluids is higher (near by 10%–20%) than water-based nanofluids with the increase in Re. This reveals that heat transfer from wall to the EG-based nanofluid is higher in comparison with water-based nanofluid. Again, the decrease in ΔT in metallic (Zn-based) nanofluids is higher (about 13%–18%) relative to metal oxide (ZnO-based) nanofluids. This indicates the higher heat transfer to metallic (Zn) nanofluid as compared to metal oxide (ZnO) nanofluid from the wall of the channel.

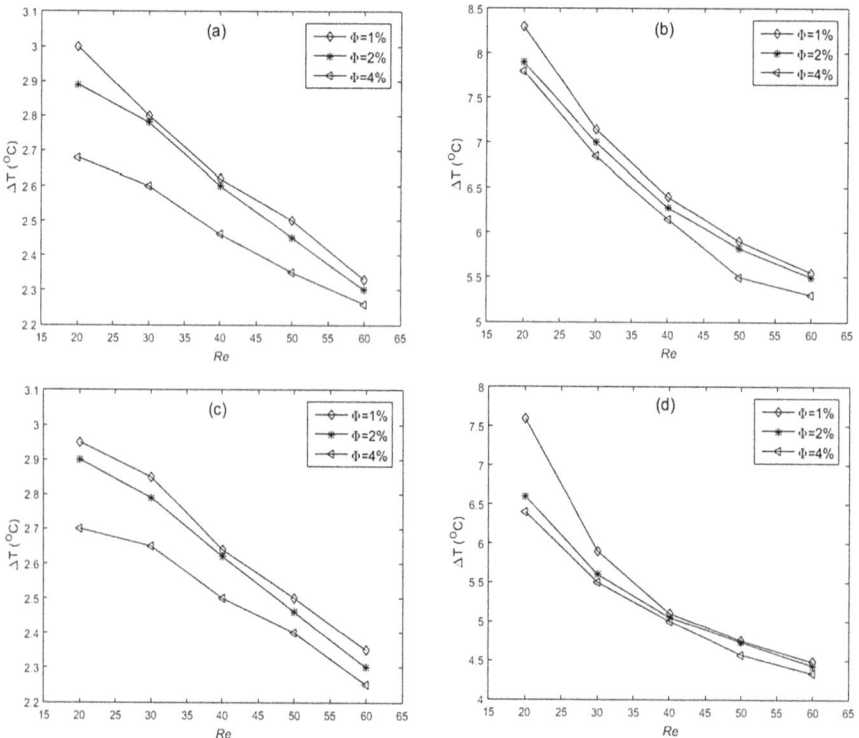

FIGURE 4.4 Influence of wall and bulk nanofluid temperature difference for (a) ZnO/W, (b) ZnO/EG, (c) Zn/W, (d) Zn/EG.

4.3.2 CONSTANT WALL BOUNDARY CONDITION

4.3.2.1 Variation of Average Heat Transfer Coefficient

Average heat transfer coefficient for the nanofluid in both the rectangular and triangular ducts can be calculated using Equation 4.18. The variation of $\bar{h}$ with Re for different volume fractions is shown in Figure 4.5. The heat transfer coefficients in both the microchannels are found to be increased with the increase in Re, and the maximum $\bar{h}$ is found when the highest volume fraction, i.e., $\phi=4\%$, is used. This trend is quite consistent with the previously published results [24]. The nature of variation of $\bar{h}$ with Re for different volume fractions is almost similar to the trend obtained by Uysel et al. [24] in rectangular microchannel using ZnO/EG nanofluid. In rectangular channel, the increase in $\bar{h}$ is about 11%–15%, whereas in triangular channel, $\bar{h}$ increases by 33%–44%. At a particular Re, for example, Re=120, the $\bar{h}$ value in the triangular channel is about 51%–70% higher than that in the rectangular channel.

4.3.2.2 Pressure Variation

The pressure drop per unit length of the microchannel using ZnO/W nanofluid for various concentrations of nanoparticle is shown in Figure 4.6. In both the microchannels, pressure drop increases with the increase in Re, and the maximum pressure drop is found to be at the highest volume fraction. The similar trend was found by other researchers in previous publications [24,25]. The pressure drop in rectangular microchannel increases by about 100%, and on the other hand, in triangular microchannel, it increases by nearly 107%, 108%. Again, at Re=120, ΔP in rectangular channels is about 11% higher than triangular channel. Although heat transfer and pressure drop increase in both the microchannels, the rate of increase of heat transfer is higher than the increase in pressure drop, and therefore, rectangular microchannel is considered for further analysis.

4.3.2.3 The Darcy Friction Factor

The influence of the Darcy friction factor of both the rectangular and triangular ducts is shown in Figure 4.7. The f at all the volume fractions in both the microchannels is

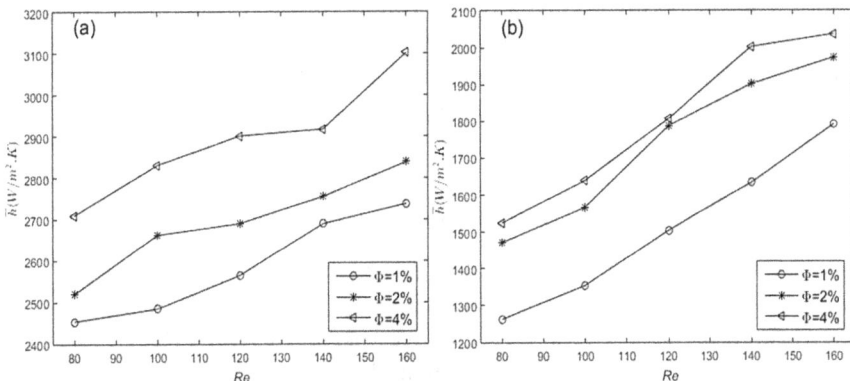

FIGURE 4.5 Variation of $\bar{h}$ of ZnO/W nanofluid in (a) rectangular and (b) triangular ducts.

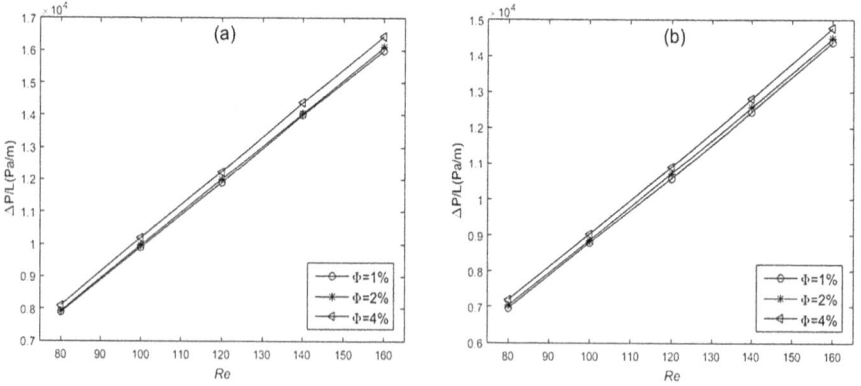

FIGURE 4.6 Pressure drop per unit length vs. Re of ZnO/W in (a) rectangular and (b) triangular microchannels.

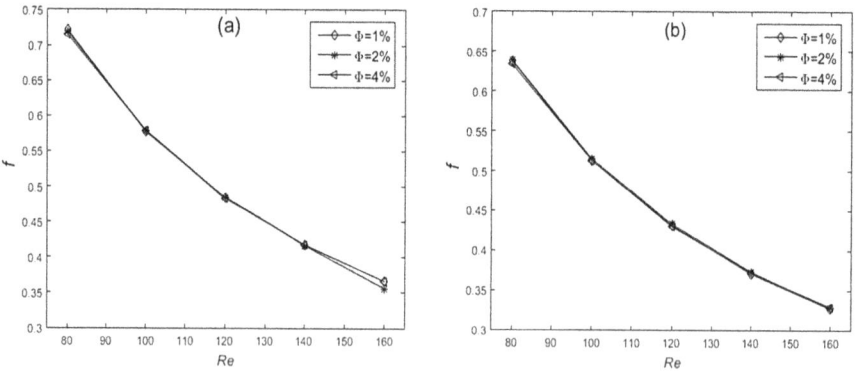

FIGURE 4.7 Influence of f with Re for ZnO/W in (a) rectangular and (b) triangular microchannels.

almost identical. However, it gradually decreases with the increase in Re. Figure 4.7 shows that f decreases by about 52% in rectangular channel, whereas about 49% decrease in the f value is found in triangular channel. Further, at a given value of Re, i.e., Re=120, f in the rectangular channel is about 12% higher than that in the triangular channel.

4.3.3   BOUNDARY CONDITION OF CONSTANT HEAT FLUX

4.3.3.1   Influence of Average Heat Transfer Coefficient

Equation 4.19 was used to calculate $\bar{h}$ of ZnO/W, ZnO/EG, Zn/W, and Zn/EG nanofluids. The variation of $\bar{h}$ with Re with three volume concentrations are shown by different plots in Figure 4.8. The plots reveal that the value of $\bar{h}$ increases with the increase in Re as well as volume fraction. This is agreed upon by other researchers also irrespective of the type of nanofluid [23]. Because of the increase in Re, the chaotic motion of the nanoparticles enhances, leading to an increase in the thermal conductivity of the nanofluid. In addition, due to the increase in volume fraction of nanoparticles, the convective effect of suspended nanoparticles increases [23]. The increase in $\bar{h}$ is about 17%–27% and 43%–69% in the case of water-based and EG-based nanofluids, respectively. From Figure 4.8a and b, it can be concluded that $\bar{h}$ value of ZnO/W nanofluid is higher (about 14%) than ZnO/EG nanofluid for all the volume fractions. However, Zn/EG nanofluid shows slightly higher value (about 4%) of $\bar{h}$ at higher Re when compared with Zn/W nanofluid. It is found that the variation of $\bar{h}$ with Re for Cu/W nanofluid is about 62% when Re increases from 300 to 850 [11], whereas in this work, similar amount of increase in $\bar{h}$ is observed at quite lower Re. Further, considering EG as the base fluid, $\bar{h}$ value is higher (about 23%) when Zn is used instead of ZnO nanoparticles (refer Figure 4.8b and d). On the other hand, metal oxide nanoparticle (ZnO) shows slightly higher rate of heat transfer in comparison with metallic nanoparticle (Zn) when water is used as the base fluid.

4.3.3.2 Pressure Variation

The variation of pressure drop with Re for ZnO/W, ZnO/EG, Zn/W, and Zn/EG nanofluids can be observed by the plots from Figure 4.9. At a particular value of

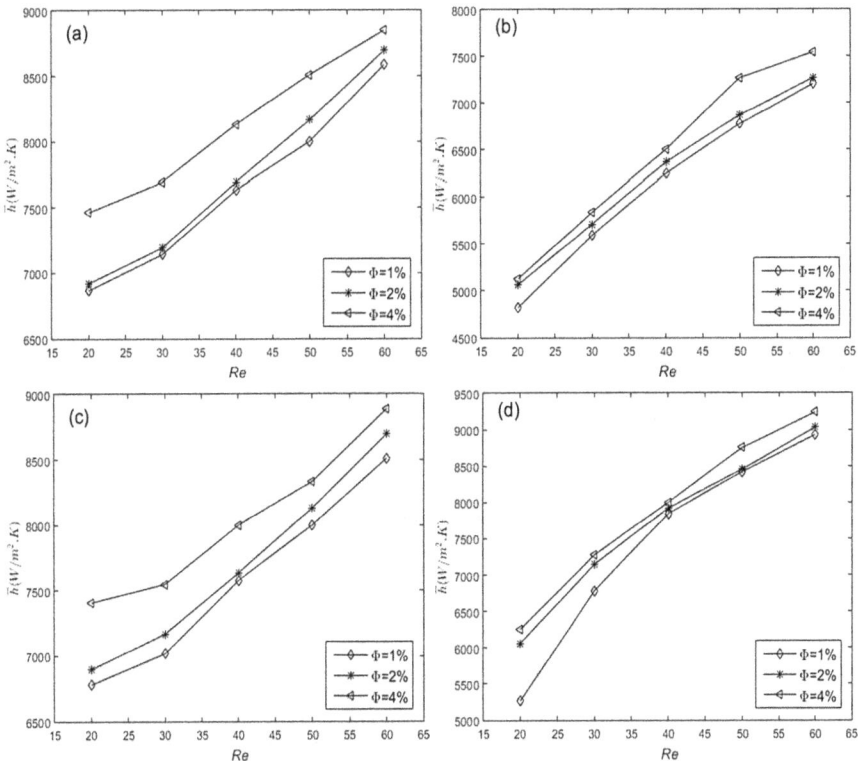

FIGURE 4.8 Variation of $\bar{h}$ with Re for (a) ZnO/W, (b) ZnO/EG, (c) Zn/W, (d) Zn/EG nanofluids in rectangular microchannel.

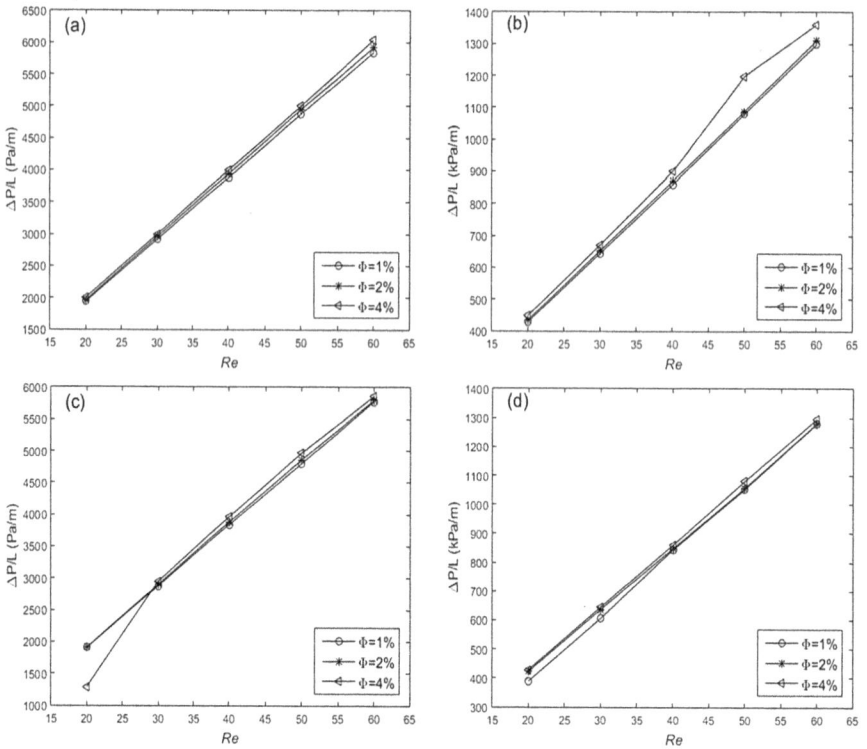

FIGURE 4.9 Pressure drop per unit length vs. Re for (a) ZnO/W, (b) ZnO/EG, (c) Zn/W, (d) Zn/EG in rectangular duct.

Re, ΔP of nanofluids is almost identical. However, the significantly higher pressure drop is observed in EG-based nanofluids when compared with water-based nanofluids. The increase in ΔP with the increase in Re of Zn/EG nanofluid is about 30% higher than ZnO/EG nanofluid. On the other hand, increase in ΔP in ZnO/W is about 125% higher than the Zn/W nanofluid. Moreover, the highest ΔP increase (i.e., about 225%) is found in Zn/EG, whereas the lowest ΔP increase (i.e., about 74%) is found to be in Zn/W nanofluid. Further, the rate of increase in pressure drop is almost identical (except slight increases at higher Re for $\phi=4\%$) with the increase in Re in all the volume fractions, whereas the rate of increase in ΔP becomes gradually higher at higher value of Re [24]. The trend of variation of ΔP with Re is quite similar with the trend obtained by Azizi et al., where the maximum increase in pressure drop was about 250% [11]. However, their research was based on Cu/W nanofluid in a microchannel heat sink of cylindrical shape at higher Re.

4.3.3.3 The Darcy Friction Factor

The Darcy friction factors for both the rectangular and triangular ducts were calculated using Equation 4.20, and they are plotted against the Re as shown in Figure 4.10. In all the nanofluids, f decreases with increase in Re, although the rate

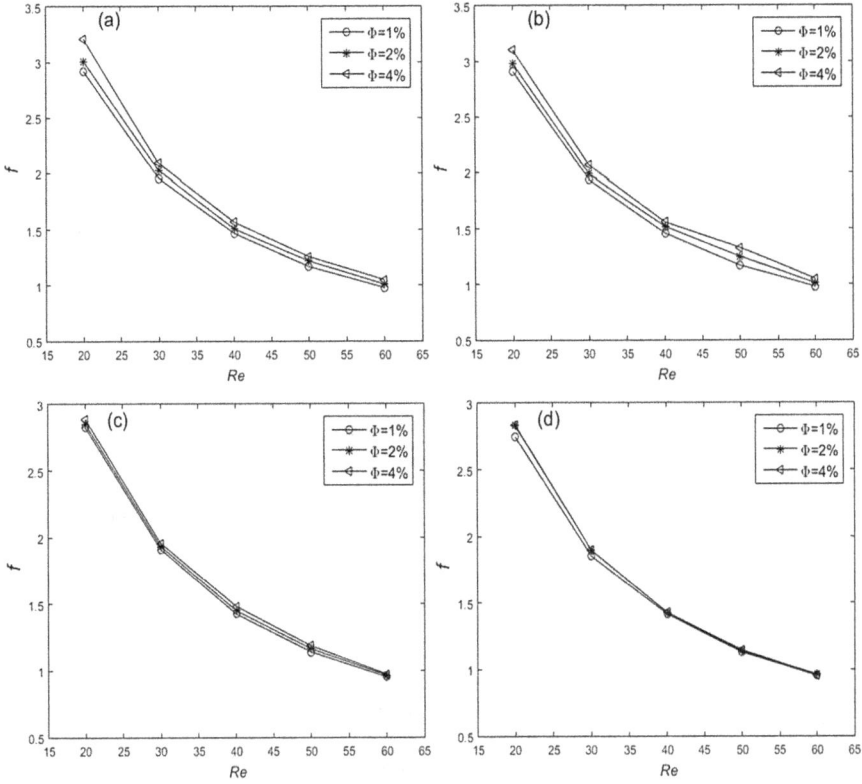

FIGURE 4.10 Influence on f of (a) ZnO/W, (b) Zn/W, (c) ZnO/EG, (d) Zn/EG nanofluids with different volume fraction.

of decreasing is very high at low Re (up to, i.e., Re=35). The similar trends are found in the microchannels of previous publications [1,2]. In particular, the Darcy friction factor of ZnO/EG (refer Figure 4.10c) for all volume fraction is approximately equal, which is quite consistent with the results obtained by Uysal et al. [24] for the same nanofluid. The metal oxide nanofluids (i.e., ZnO/W and ZnO/EG) exhibit higher f (about 10%) than metallic nanofluids (i.e., Zn/W and Zn/EG). It is observed that varying the volume fraction of nanoparticles does not significantly affect f. However, the nanofluids with higher volume fraction (e.g., $\phi=4\%$) show slightly higher value of f (about 8%–10%) in metal oxide nanofluids at low Re.

4.3.4 INFLUENCE OF VOLUME FRACTION ON $\bar{h}$ AND ΔP

The influence of volume fraction on heat transfer at Re=20 shown in Figure 4.11a indicates that the heat transfer coefficients of all water-based nanofluids are almost similar. However, they are quite higher than EG-based nanofluids (about 15% than Zn/EG and 37% than ZnO/EG). Further, EG-based metallic nanofluid (Zn/EG) shows significantly higher (about 25%) $\bar{h}$ value than metal oxide nanofluid (ZnO/

EG) at high volume fraction (e.g., $\phi=4\%$). On the other hand, this difference is lower (about 9% only) at small volume fraction such as, $\phi=1\%$. In all the nanofluids, only a small amount (2%–10%) of $\bar{h}$ value is found to be increased with the increase in volume fractions. Figure 4.11b shows that by increasing volume fraction of nanoparticles in base fluid, pressure drop is hardly affected except an increase of about 3%–5% in water-based nanofluids. Again, pressure drop in water-based nanofluids is quite higher (328%–364%) in comparison with EG-based nanofluids in all the volume fractions. In water-based nanofluids, although the rate of heat transfer (i.e., $\bar{h}$) is higher (about 15%–37%), the pressure drop is also significantly higher in comparison with EG-based nanofluids.

4.3.5 INFLUENCE OF MICROCHANNEL GEOMETRY ON $\bar{h}$ AND ΔP

Figure 4.12 shows that $\bar{h}$ and ΔP are increased by increasing the volume fraction of nanoparticle in both triangular and rectangular microchannels. The increase in

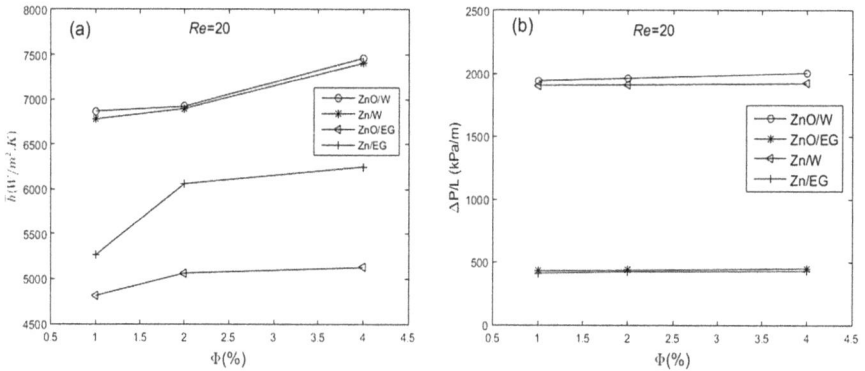

FIGURE 4.11 Influence of (a) $\bar{h}$ and (b) ΔP of the nanofluid with ϕ.

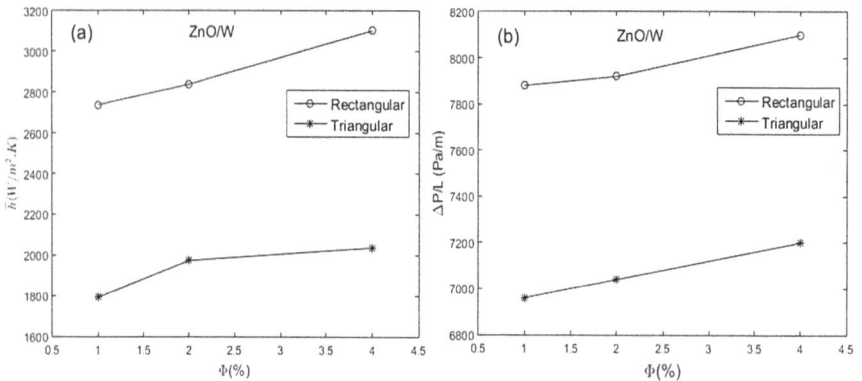

FIGURE 4.12 Influence of (a) $\bar{h}$ and (b) ΔP of the nanofluids with ϕ at different geometrics.

$\bar{h}$ from 11% to 13% in both rectangular and triangular microchannels when the ϕ changes from 1% to 4% (refer Figure 4.12a). On the other hand, ΔP increases with volume fraction are about 3% and 4% in rectangular and triangular microchannels, respectively (Figure 4.12b). However, the $\bar{h}$ value in rectangular microchannel is about 53%–55% higher with respect to triangular microchannel. Similarly, pressure drop in rectangular duct is about 12%–14% higher than the triangular duct.

4.4 CONCLUSIONS

The difference of temperature between wall and nanofluid decreases with the increase in Re. The decrease in ΔT in EG-based nanofluids is 10%–20% higher than water-based nanofluid, whereas the decrease in ΔT in Zn-mixed nanofluid is 13%–18% higher than ZnO-mixed nanofluid. At constant wall boundary condition, $\bar{h}$ increases by 11%–15% in rectangular and 33%–44% in triangular microchannel with the increase in Re. However, for a given Re, $\bar{h}$ of rectangular channel is about 51%–70% and ΔP is about 11% higher than triangular channel. When the wall temperature remains constant, f decreases with Re and for a particular value of Re, it is about 12% higher in rectangular channel than triangular channel. At constant heat flux boundary condition, $\bar{h}$ is higher in ZnO/W than ZnO/EG and in Zn/EG than Zn/W nanofluid. Again, $\bar{h}$ is higher in Zn/EG than ZnO/EG, and ZnO/W has higher value compared to Zn/W nanofluid. In this condition, the increase in ΔP of Zn/EG is 30% higher than ZnO/EG, whereas in ZnO/W, ΔP is about 125% higher than Zn/W nanofluid. The Darcy friction factor decreases with Re, and metallic oxide nanofluids exhibit about 10% higher f than metallic nanofluids. Only 2%–10% increase in $\bar{h}$ and 3%–5% increase in ΔP are found with the increase in ϕ. Moreover, ΔP is about 328%–364% higher in water-based nanofluid in comparison with EG-based nanofluid. Further, $\bar{h}$ and ΔP in rectangular microchannel are about 53%–55% and 12%–14% higher than triangular microchannel, respectively.

REFERENCES

1. Mala, G. M. and Li, D. (1999). Flow characteristics of water in microtubes. *International Journal of Heat and Fluid Flow*, 20:142–148.
2. Jung, J.-Y. and Kwak, H.-Y. (2008). Fluid flow and heat transfer in microchannels with rectangular cross section. *Heat Mass Transfer*, 44:1041–1049.
3. Tuckerman, D. B. and Pease, R. F. W. (1981). High-performance heat sinking for vlsi. *IEEE Electron Device Letters*, 2(5):126–129.
4. Ting, H.-H. and Hou, S. S. (2016). Numerical study of laminar flow and convective heat transfer utilizing nanofluids in equilateral triangular ducts with constant heat flux. *Materials*, 9(7):576.
5. Behnampour, A., Akbari, O. A., Safaei, M. R., Ghavami, M., Marzban, A., Shabani, G. A. S. and Mashayekhi, R. (2017). Analysis of heat transfer and nanofluid fluid flow in microchannels with trapezoidal, rectangular and triangular shaped ribs. *Physica E: Low-Dimensional Systems and Nanostructures*, 91:15–31.

 6. Zeinali Heris, S., Kazemi-Beydokhti, A., Noie, S. and Rezvan, S. (2012). Numerical study on convective heat transfer of Al$_2$O$_3$/water, cuo/water and cu/water nanofluids through square cross-section duct in laminar flow. *Engineering Applications of Computational Fluid Mechanics*, 6(1):1–14.
 7. Khaled, A.-R. and Vafai, K. (2005). Heat transfer enhancement through control of thermal dispersion effects. *International Journal of Heat and Mass Transfer*, 48(11):2172–2185.
 8. Lotfi, R., Saboohi, Y. and Rashidi, A. (2010). Numerical study of forced convective heat transfer of nanofluids: comparison of different approaches. *International Communications in Heat and Mass Transfer*, 37(1):74–78.
 9. Ahmed, H., Mohammed, H. and Yusoff, M. Z. (2012). Heat transfer enhancement of laminar nanofluids flow in a triangular duct using vortex generator. *Superlattices and Microstructures*, 52(3):398–415.
10. He, Y., Shao, B. and Cheng, H. (2014). Numerical simulation and size optimization of rectangular micro-channel heat sinks. *Appl. Math. Mech*, 35(3):278–286.
11. Azizi, Z., Alamdari, A. and Malayeri, M. R. (2015). Convective heat transfer of cu–water nanofluid in a cylindrical microchannel heat sink. *Energy conversion and management*, 101:515–524.
12. Rimbault, B., Nguyen, C. T. and Galanis, N. (2014). Experimental investigation of cuo–water nanofluid flow and heat transfer inside a microchannel heat sink. *International Journal of Thermal Sciences*, 84:275– 292.
13. Kalteh, M. (2013). Investigating the effect of various nanoparticle and base liquid types on the nanofluids heat and fluid flow in a microchannel. *Applied Mathematical Modelling*, 37(18–19):8600–8609.
14. Vajjha, R. S. and Das, D. K. (2009). Specific heat measurement of three nanofluids and development of new correlations. *Journal of Heat Transfer*, 131(7):071601.
15. Satti, J. R., Das, D. K. and Ray, D. (2016). Specific heat measurements of five different propylene glycol based nanofluids and development of a new correlation. *International Journal of Heat and Mass Transfer*, 94:343–353.
16. Esfe, M. H. and Saedodin, S. (2014). An experimental investigation and new correlation of viscosity of ZnO–Eg nanofluid at various temperatures and different solid volume fractions. *Experimental Thermal and Fluid Science*, 55:1–5.
17. Esfe, M.H. and Saedodin, S. (2014). Experimental investigation and proposed correlations for temperature dependent thermal conductivity enhancement of ethylene glycol based nanofluid containing ZnO nanoparticles. *Journal of Heat and Mass Transfer Research (JHMTR)*, 1(1):47–54.
18. Yu, W., Xie, H., Chen, L. and Li, Y. (2009). Investigation of thermal conductivity and viscosity of ethylene glycol based zno nanofluid. *Thermochimica Acta*, 491(1–2):92–96.
19. Prajapati, O., Rohatgi, N. and Rajvanshi, A. (2013). Heat transfer behaviour of nano fluid at high pressure. *Journal of Materials Science and Surface Engineering*, 1(1):1–3.
20. Salman, B., Mohammed, H. A., Munisamy, K. M. and Kherbeet, A. S. (2015). Three-dimensional numerical investigation of nanofluids flow in micro- tube with different values of heat flux. *Heat Transfer—Asian Research*, 44(7):599–619.
21. Hosseini, S. R., Sheikholeslami, M., Ghasemian, M. and Ganji, D. D. (2018). Nanofluid heat transfer analysis in a microchannel heat sink (MCHS) under the effect of magnetic field by means of KKL model. *Powder Technology*, 324:36–47.
22. Mohammad, Z. and Jalel, A. (2018). Heat transfer analysis of nanofluid based microchannel heat sink. *International Journal of Heat and Mass Transfer*, 127(Part B);1233–1242.
23. Kahani, M. (2019). Simulation of nanofluid flow through rectangular microchannel by modified thermal dispersion model. *Heat Transfer Engineering*, 1521–1537. https://doi.org/10.1080/01457632.2018.1540464.

24. Uysal, C., Arslan, K. and Kurt, H. (2016). A numerical analysis of fluid flow and heat transfer characteristics of Zno-Ethylene glycol nanofluid in rectangular microchannels. *Strojniški vestnik-Journal of Mechanical Engineering*, 62(10):603–613.

25. Bowers, J., Cao, H., Qiao, G., Li, Q., Zhang, G., Mura, E. and Ding, Y. (2018). Flow and heat transfer behaviour of nanofluids in microchannels. *Progress in Natural Science: Materials International*, 28: 225–234

5 Influence of Layer Thickness and Orientation on Tensile Strength and Surface Roughness of FDM-Built Parts

Praveen Kumar Nayak, Anshuman Kumar Sahu, and Siba Sankar Mahapatra
National Institute of Technology Rourkela

CONTENTS

5.1 INTRODUCTION

Additive manufacturing (AM) technologies provide design freedom to designers and engineers to bring mechanical design to life. They also enable customization of products based on customer-driven requirements. AM technologies like fused deposition modeling (FDM), stereolithography (SL), laminated object manufacturing (LOM), and direct metal laser sintering (DMLS) have seen larger implementation, as industries move towards smarter manufacturing techniques. The FDM process involves the extrusion of semi-molten thermoplastic polymer filament through an extruder nozzle which is controlled by numeric control software. The material is deposited layer by layer to form complex three-dimensional shapes. FDM processes facilitate a rapid iterative testing of design concepts and can serve as a rapid manufacturing tool for very short runs. But the freedom of the FDM processes comes with shortcomings as well in terms of accuracy, strength, surface finish, and productivity. Due to the relatively recent development of the FDM processes, the mechanical properties of the FDM parts are still being rigorously studied and validated before the processes gain wider adoption in the mainstream production process. The FDM processes are influenced by a number of process parameters, and minor variations during the fabrication process can result in drastically different mechanical properties and dimensional attributes of the final product. There is a need to establish industrial standards to maintain the quality of parts as well as for testing and characterizing the part properties.

In this study, an effort has been made to demonstrate a relationship between the layer thicknesses and build orientation during the fabrication process and the tensile strength and surface roughness of the final part. The tensile properties and surface roughness have been studied, and the surface morphology has been observed for the fabricated FDM tensile specimens. Further, the process parameters have been optimized using multiobjective optimization on the basis of ratio analysis (MOORA) method. The significant factors contributing to the tensile properties and the surface roughness have been identified, and the optimum parameters have been selected for obtaining the best possible parts.

5.2   LITERATURE REVIEW

Recently, the FDM process has acquired attention by researchers. Some of the studies carried out in the field of FDM process is discussed as follows. Rodriguez et al. found a significant correlation between the mesostructure of the filament and the stress–strain response while studying the mechanical behavior of acrylonitrile butadiene styrene (ABS) FDM parts. The study also suggested that the voids between layers during extrusion process also contribute to the decrease in FDM parts [1]. Ahn et al. examined the effect of process parameters on the building of ABS P400 parts and found that the raster orientation and air gap has significant influence on the mechanical strength of the parts. The study found that FDM ABS parts exhibit mechanical properties which are inferior compared to the injection molding parts. It also suggested basic building rules for building of the FDM parts [2]. Sun et al. investigated the mechanisms controlling the formation of the bonds between the layers of extruded filaments in the FDM process. The temperature profile of the adjoining polymer filaments was studied under different processing conditions, and the related mechanical properties were observed. The study showed that the building strategy and the build envelope conditions had a significant effect on the cooling temperature profile as well as the degree of coalescence between the layers [3]. Sood et al. used response surface methodology to study the effect of layer thickness, build orientation, raster angle, and raster width and air gap on the FDM parts. The tensile, flexural, and impact strength were found to be dependent on the number of layers in the part. A desirability function approach had been used to determine the optimal control factor settings for the process [4]. Luzanin et al. explored the relation between process parameters and flexural strength with special attention given to infill percentages and found a high degree of non-linearity with inputs and effects with the extrusion speed, deposit angle, and infill exhibiting significant influence on the flexural strength [5]. Durgun and Ertan suggested in their study that the build orientation has more significance than raster angle on the FDM process while testing for surface roughness, tensile strength, and flexural strength. The surface roughness and tensile strength were found to be interdependent, and the horizontal build orientation was found to be optimal [6]. Hedayati and Zadpoor explored the analytical relationships which can be used to compare the mechanical properties of additively manufactured porous biomaterials by simplifying the structures into different types of unit cells [7]. Anitha et al. used the Taguchi technique to determine the critical process parameters which influence the surface quality of the parts produced by the FDM process and found the layer thickness to be most significant contributor to the surface roughness of the parts [8]. Ahn et al. suggested a mathematical expression to express the surface roughness distribution according to changes in the surface angle. The expression was obtained by taking the main factors contributing to the surface quality as well as the geometry of the deposited filament into consideration, and the empirical data was found to be in close agreement with actual surface roughness data obtained from actual FDM specimen [9]. Galantucci et al. suggested the surface finish of FDM parts improved on subjecting the parts to chemical finishing which significantly reduced the surface roughness when compared to the untreated specimens. The suggested treatment process was also relatively inexpensive and required little human intervention [10]. Sood et al. explored the dependency of compressive stress on process parameters with the study revealing a high degree of non-linearity between the input and the response revealing the high degree of complexity in the relationship between the compressive strength and the process parameters. A statistically validated predictive equation was also developed and the computed data was compared against data predicted using artificial neural network. The proposed equation provided a reliable method of predicting and optimizing the FDM process and mechanical behavior for future applications [11]. Applications of FDM to fabricate components and parts for different industrial sections were explored in the past studies, revealing these applications have great potential [12–14].

5.3 MATERIALS AND METHODS

The material used for the test specimen fabrication is ABS M30. ABS (chemical formula $((C_8H_8 \cdot C_4H_6 \cdot C_3H_3N)n)$ is a common thermoplastic. It is a thermoplastic polymer and is commonly used for making a wide range of plastic products. The monomer of ABS is shown in Figure 5.1. It is a terpolymer which is formed by the polymerization of acrylonitrile and styrene in the presence of butadiene. Its properties such as high hardness and impact strength, rigidity, high toughness, high chemical and thermal resistance, resistance to distortion from humidity, negligible creep, and easy processing with less cost increased its use in aerospace, automobile, electronics, and process industries. ABS is resistant to aqueous acids and alkalis, and has fairly constant electrical and mechanical properties over the acceptable range of operating temperature and humidity.

FDM parts are built from a three-dimensional computer-generated model that describes the internal and external geometry of the part. The 3D model is usually prepared using a conventional CAD software or by using a 3D scanner to scan an

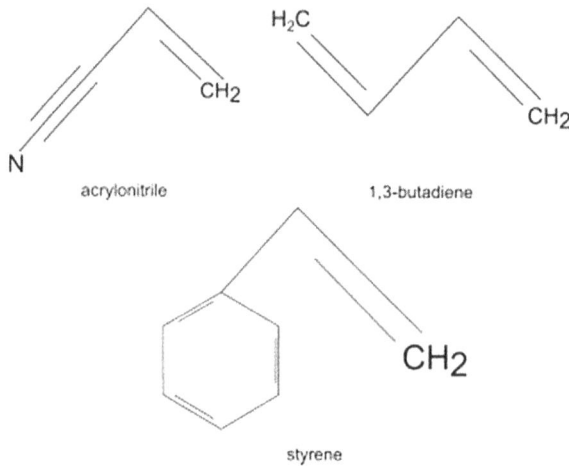

H₂C

CH2

acrylonitrile

1,3-butadiene

CH2

CH2

styrene

FIGURE 5.1   Monomers of ABS.

existing part. STL format is a common format which is used for most of the FDM systems, and most of the CAD models are be converted into STL format before being processed on the FDM system. This format generates a tessellated model of the CAD model, which can further be processed for the calculation of the slices. The STL file is loaded onto processing software which can be used for basic model manipulation as well as subsequent clean-up of the digital model. The machine setup includes loading of the material, preheating of build chamber, preparation of the build plat-form, and other related activities before the build process. The actual generation of the part is highly automated with very little human intervention. Some monitoring of the machine is required to ensure the prevention of errors such as running out of material, dislocation of the part during the building process, software glitches, and power outage. The FDM process is greatly influenced by numerous process param-eters such as layer thickness, raster orientation, air gap, and part orientation, and these process parameters have a significant influence on the mechanical and physical characteristics of the FDM parts. Figure 5.2 gives a visual representation of the key process parameters of the FDM process. Some of the key process parameters are described as follows:

a. Part interior style:
   Controls the manner in which the interior raster is deposited. It is of three types:
   i.   Solid: 100% infill which leads to a fully dense part leading to higher cost and lower speeds.
   ii. Sparse – high density: 70% infill which leads to lower cost and moder-ate build speed. This setting generally used for cost and weight savings while maintaining good mechanical strength.
   iii. Sparse – low density: 30% infill which leads to the lowest cost with much higher build speed. This setting is used for fast prototyping and rapid development when mechanical strength is not very important.

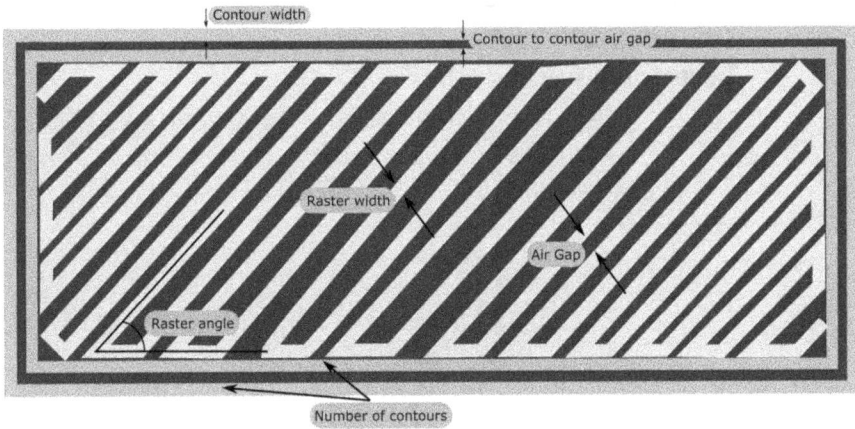

FIGURE 5.2 Representation of the cross section of a layer during FDM process with significant features.

b. Part fill style

Controls the style in which the part will be filled. The default option of "Normal Rasters" is used in this study.

c. Raster angle

It is the direction of the raster relative to the x-axis of the build platform. The raster rotates by 90° every successive layer to reduce anisotropy.

d. Layer thickness

It is the thickness of the layer deposited and is dependent on the diameter of the nozzle used.

e. Raster width

It is the width of the raster used in the interior fill style.

f. Raster gap

It is also termed as "air gap". It is the gap between two adjacent layers as well as the gap between the perimeter contour and the adjacent raster.

g. Contour width

It is the width of contour which surrounds the infill and creates the edges of the layer. Each layer has at least one contour.

h. Part build orientation

It refers to the orientation of the part on the build platform with respect to the x, y, and z axes with the x and y axes describing the horizontal plane of the build platform and the z axis the vertical direction with respect to the build platform.

It has been found from studies that part orientation and layer thickness have a significant influence on the tensile characteristics as well as the surface morphology of the parts. Table 5.1 shows the factors to be studied and the selected levels for the factors.

The mechanical strength of ABS-built parts is determined by conducting tensile tests according to ASTM D638 standard test method for tensile properties of

TABLE 5.1

Factors and Their Levels for the FDM Parts

Symbol	Factors	Unit	Level 1	Level 2	Level 3
A	Layer thickness	mm	0.178	0.254	0.330
B	Build orientation	°	0	15	30

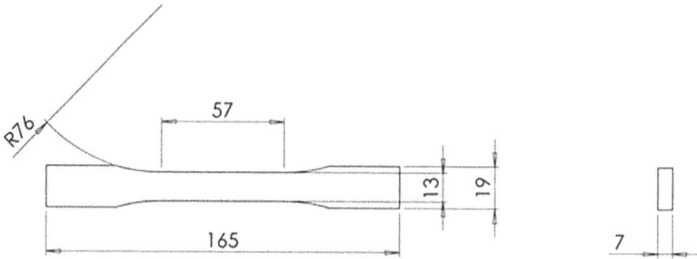

FIGURE 5.3   Two-dimensional CAD model for FDM tensile test specimen with dimensions in millimeter.

plastics. The tensile specimens are designed using SOLIDWORKS 2014 as shown in Figure 5.3.

The CAD model is exported to an STL format which is imported onto a FORTUS 400MC FDM system. The machine as shown in Figure 5.4 is manufactured by Stratasys Inc. and has a build volume of 355 mm × 254 mm × 254 mm. The system supports various types of build materials such as ABS M30, PC-ABS, PC, and ULTEM as well as support materials such as ABSi, ABS-ESD7, and PC-ISO. It has one build material bay and one support material bay which can hold material canisters of 92 in³. The operating room temperature is controlled to be less than 29.4°C. The material used during the fabrication of test specimen on FORTUS 400MC is ABS M30. The material is extruded from the extruder nozzle at a temperature of 230°C.

The STL file is imported onto the INSIGHT software package for processing and the part model onto the machine. The layer thickness and build orientation of the parts as shown in Figure 5.5 are defined inside the Insight software package, and the 3D model is broken into individual slices followed by the generation of tool paths.

The nozzle tip of the build material is varied to vary the layer thickness, whereas the nozzle tip of the support material is kept the same throughout the experiment. The raster angle is rotated by 90° every consecutive layers to obtain better interlocking between layers which lead to a reduction in the anisotropic characteristics of the FDM process. The test specimen is removed from the build platform and placed in a solvent solution to remove the dissolvable supports from the test specimen. The built test specimens are shown in Figure 5.6.

Experiments have been carried out based on a full factorial experimental design to determine the effect of process parameters on the FDM-built parts. The tensile

FIGURE 5.4   Stratasys FORTUS 400MC rapid prototyping system.

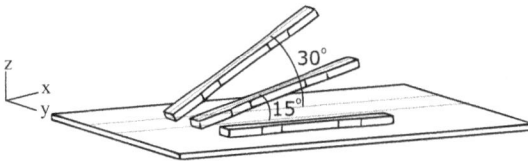

FIGURE 5.5   Representation of different build orientations of the FDM specimen on the build platform.

testing of the built test specimens has been carried out on a SATEC 600 KN Universal Testing Machine with cross-head moving at 1 mm/min to provide a tensile load on the test specimen at ambient temperature until fracture of the test specimen. The average surface roughness of the test specimen for different orientations has been obtained using an instrument Taylor Hobson Talysurf with the cutoff length set at 25 mm and with two steps. The surface morphology of the fabricated tensile specimen has been studied under a scanning electron microscope.

5.4   RESULTS AND DISCUSSION

The ABS parts are built by the FDM process by varying the input parameter layer thickness and build orientation as presented in Table 5.1. The experimental results like ultimate tensile strength (UTS) and surface roughness obtained from the test result recorded for each specimen are shown in Table 5.2.

Analysis of variance (ANOVA) study is carried for the UTS as well as the surface roughness (R_a) to identify the factors which have a significant influence on the

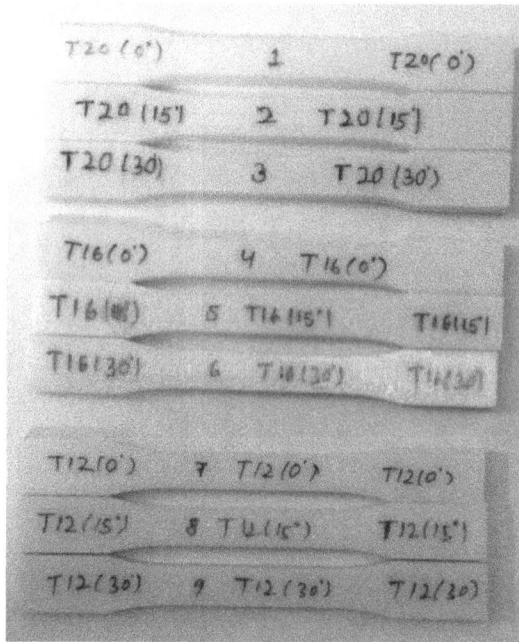

FIGURE 5.6 Tensile specimen fabricated on Fortus 400MC machine.

TABLE 5.2
Results Obtained from Experimental Runs

Experiment Number	A	B	UTS (MPa)	R_a (µm)
1	1	1	43.07	7.77
2	1	2	41.25	38.11
3	1	3	39.83	32.5
4	2	1	48.68	9.08
5	2	2	45.24	40.02
6	2	3	42.5	37.45
7	3	1	53.1	10.32
8	3	2	48.25	45.21
9	3	3	45.24	41.5

response. The R^2 values in ANOVA for tensile strength and surface roughness are 96.0% and 99.3%, respectively. The ANOVA study for the UTS and the R_a is given in Tables 5.3 and 5.4, respectively. If the p-value of the ANOVA is less than 0.05, then the effect of the input parameter on the output result is said to be significant. It is observed that both layer thickness and build orientation have a significant influence on both the output responses. The layer thickness is a more significant factor for the

TABLE 5.3
ANOVA for Tensile Strength

Source	DF	Adj SS	Adj MS	F-Value	P-Value
Layer thickness (A)	2	84.111	42.055	29.7	0.004*
Build orientation (B)	2	50.273	25.136	17.75	0.010*
Error	4	5.663	1.416		
Total	8	140.047			

*Significant at 95% confidence interval.
$S=1.190$, $R^2=96.0\%$, $R^2(\text{adj})=91.9\%$.

TABLE 5.4
ANOVA for Surface Roughness

Source	DF	Adj SS	Adj MS	F-Value	P-Value
Layer thickness (A)	2	58.27	29.13	9.24	0.032*
Build orientation (B)	2	1832.57	916.287	290.46	0.000*
Error	4	12.62	3.155		
Total	8	1903.46			

*Significant at 95% confidence interval.
$S=1.776$, $R^2=99.3\%$, $R^2(\text{adj})=98.7\%$.

tensile strength of the part compared to the build orientation. But the surface roughness is more dependent on the build orientation as compared to the layer thickness.

The optimum parameters are obtained from the main effects plot shown in Figure 5.7. It is observed that the tensile specimen built at a layer thickness of 0.330 mm and a built orientation of 0° has the greatest tensile strength. It is also observed that the tensile specimen built at a layer thickness of 0.330 mm and a build orientation of 15° has the least surface roughness.

It is observed from the main effects plot that the tensile strength of parts increases with an increase in the layer thickness but decreases with an increase in the build orientation. The tensile strength is greatly influenced by the degree of cohesion and the thermal gradient between the layers of the build part. The layer thickness and the build orientation have a direct correlation with the degree of cohesion and the thermal gradient. The thermal gradient and the degree of cohesion are dependent on the time taken to build the parts. The time taken to build the part reduces with an increase in the layer thickness, whereas the build time increases with an increase in the build orientation. The thermal gradient and the degree of cohesion between the layers give rise to defects such as internal cracks and porosities which are shown in Figure 5.8. The defects lead to the variation in the tensile strength of the FDM-built part. The fracture specimens of the ABS parts after the UTS test are shown in Figure 5.9.

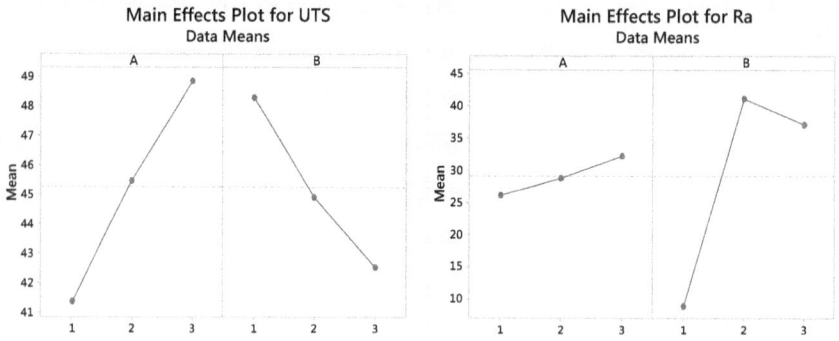

FIGURE 5.7   Main effects plot for UTS and average surface roughness.

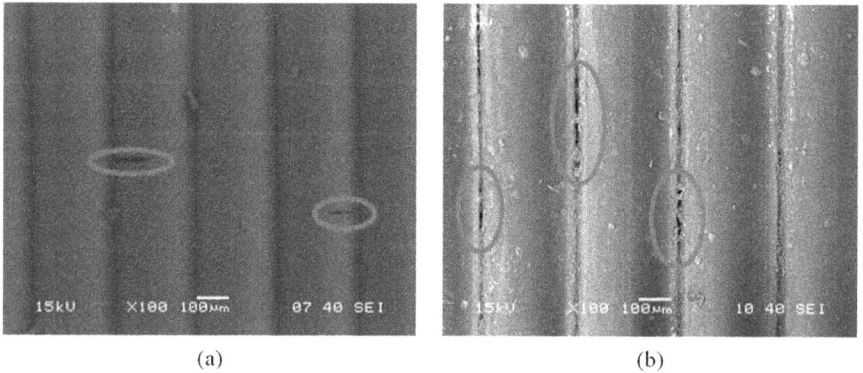

FIGURE 5.8   Internal defects observed in the tensile specimen built at a build orientation of 0° and a layer thickness of 0.254 mm with (a) showing internal cracks and (b) showing interstitial porosities.

It is also observed that the surface roughness increases linearly with an increase in the layer thickness. But the surface roughness does not increase linearly with the increase in the build orientation. The parts built at a build orientation of 0° has the least surface roughness, and the surface roughness increases when the build orientation is increased. But, the surface roughness of the parts built at a build orientation of 30° is lower than that of the parts built at a build orientation of 15°. This is due to the staircase effect of the FDM process (Figure 5.10). The effect of the staircasing of the layers reduces as the build orientation is increased which leads to better surface finish when build orientation is increased from 15° to 30°.

5.5   OPTIMIZATION BY MOORA METHOD

It is observed that the optimum parameters are different for the tensile strength and surface roughness. But there is a need to obtain good tensile strength at lower surface

(a) (b)

(c)

FIGURE 5.9 Fractured specimen built at a layer thickness of 0.254 mm and at (a) a build orientation of 0°, (b) 150°, (c) 300°.

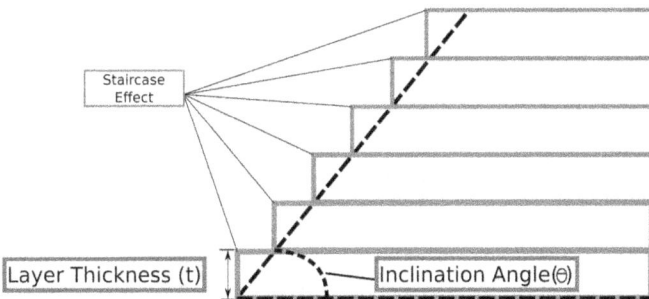

FIGURE 5.10 Representation of staircase effect in FDM-built parts.

roughness. So, multicriteria decision-making is utilized to obtain the optimum parameter setting which would provide suitable results. Here, the MOORA method is used to find out the optimum parameters for the FDM process, and it has been used to convert multi-responses into single performance index. Two responses are considered here simultaneously. The MOORA index is calculated as follows [15–17]:

1. Calculate the normalized value:

$$X_{ij}^* = \frac{X_{ij}}{\sqrt{\sum_{i=1}^{m} X_{ij}^2}} \quad (j = 1, 2, \ldots, n)$$

(5.1)

where, X_{ij} is the observed responses and X_{ij}^* normalized value.

2. Calculate the MOORA index:

$$y_i = \sum_{j=1}^{m} X_{ij}^* - \sum_{j=m+1}^{m} X_{ij}^*$$

(5.2)

$$y_i = \sum_{j=1}^{m} w_j X_{ij}^* - \sum_{j=m+1}^{m} w_j X_{ij}^*$$

(5.3)

where y_i is the normalized assessment value or the MOORA index, m maximized number of responses, $n - m$ minimized number of responses, and w_j weight of the jth response.

The normalized value is added in case of maximization problem and subtracted in case of minimization problem.

In this FDM process, UTS is considered as maximization response, and average surface roughness is considered as minimization response.

The calculated normalized value for individual responses and the MOORA index are shown in Table 5.5. ANOVA is carried out for the MOORA index to find the significant parameters. The results are shown in Table 5.6. ANOVA is carried out to determine the significant factors contributing to the MOORA index. Table 5.6 shows the results of the analysis.

TABLE 5.5
Result Obtained from Gray Relational Analysis of the FDM Process

Ex. No.	A	B	UTS	R_a	x_{ij}^* (UTS)	x_{ij}^* (R_a)	y_i
1	1	1	43.07	7.77	0.0570	0.0796	−0.0113
2	1	2	41.25	38.11	0.2797	0.3904	−0.0553
3	1	3	39.83	32.5	0.2386	0.3329	−0.0472
4	2	1	48.68	9.08	0.0666	0.0930	−0.0132
5	2	2	45.24	40.02	0.2938	0.4100	−0.0581
6	2	3	42.57	37.45	0.2749	0.3837	−0.0544
7	3	1	53.13	10.32	0.0758	0.1057	−0.0150
8	3	2	48.25	45.21	0.3319	0.4632	−0.0657
9	3	3	45.24	41.5	0.3046	0.4251	−0.0603

TABLE 5.6

ANOVA Table for the MOORA Index

Source	DF	Adj SS	Adj MS	F-Value	P-Value
A	2	0.000123	0.000061	9.24	0.032*
B	2	0.003864	0.001932	290.46	0.000*
Error	4	0.000027	0.000007		
Total	8	0.004014			

*Significant at 95% confidence interval.
$S=0.002579$, $R^2=99.3\%$, $R^2(\text{adj})=98.7\%$.

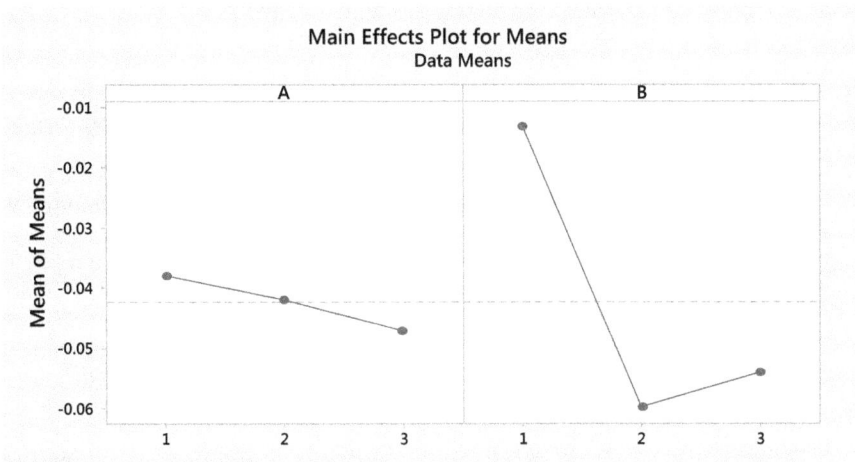

FIGURE 5.11 Main effects plot for the MOORA index.

It is observed from the ANOVA table that both the process parameters, i.e., layer thickness and build orientation, have a significant influence on the MOORA index. So, both the parameters are significant for improving the tensile strength and surface finish of the FDM-built parts. The optimum parametric settings for the FDM process are obtained from the main effects plot for the MOORA index as shown in Figure 5.11. A layer thickness of 0.178 mm and a build orientation of 0° is found to be the optimum parameter setting for obtaining better tensile strength while reducing surface roughness at the same time.

5.6 CONCLUSIONS

The study of AM processes like FDM in particular is gaining significance with an increasing need of AM technologies in various sectors. Mechanical strength of FDM parts and surface finish of the finished product are important criteria which will define greater adoption in the industry. This study has been carried out aiming to identify critical parameters which will affect the abovementioned criteria. It is found both layer thickness and build orientation are significant contributors which dictate the surface roughness as well as the tensile strength of the FDM-built parts. It is found that a layer thickness of 0.330 mm and a build orientation of 0° are the optimum parameters for maximizing tensile strength, and a layer thickness of 0.330 mm and a build orientation of 15°are optimum for minimizing surface roughness. But the optimum parameters for maximizing tensile strength while minimizing surface roughness at the same time are obtained by the MOORA method and found to be a layer thickness of 0.178 mm and a build orientation of 0°. Further research is required which would take a wider range of data while defining factors and their levels. Future studies should consider different factors such as raster width, raster angle, and air gap while optimizing the process for better understanding and utilization of the FDM process.

REFERENCES

1. Rodríguez JF, Thomas JP, Renaud JE (2001) Mechanical behaviour of acrylonitrile butadiene styrene (ABS) fused deposition materials, Experimental investigation. *Rapid Prototyping Journal*, 7(3):148–158.
2. Ahn SH, Montero M, Odell D, Roundy S, Wright PK (2002) Anisotropic material properties of fused deposition modeling ABS. *Rapid Prototyping Journal*, 8(4): 248–257.
3. Sun Q, Rizvi GM, Bellehumeur CT, Gu P (2008) Effect of processing conditions on the bonding quality of FDM polymer filaments. *Rapid Prototyping Journal*, 14(2):72–80.
4. Sood AK, Ohdar RK, Mahapatra SS (2010) Parametric appraisal of mechanical property of fused deposition modelling processed parts. *Materials & Design*, 31(1):287–295.
5. Lužanin O, Movrin D, Plančak M (2014) Effect of layer thickness, deposition angle, and infill on maximum flexural force in FDM-built specimens. *Journal for Technology of Plasticity*, 39(1):49–58.
6. Durgun I, Ertan R (2014) Experimental investigation of FDM process for improvement of mechanical properties and production cost. *Rapid Prototyping Journal*, 20(3):228–235.
7. Hedayati R, Sadighi M, Mohammadi-Aghdam M, Zadpoor AA (2017) Analytical relationships for the mechanical properties of additively manufactured porous biomaterials based on octahedral unit cells. *Applied Mathematical Modelling*, 46:408–422.
8. Anitha R, Arunachalam S, Radhakrishnan P (2001) Critical parameters influencing the quality of prototypes in fused deposition modelling. *Journal of Materials Processing Technology*, 118(1–3):385–388.
9. Ahn D, Kweon JH, Kwon S, Song J, Lee S (2009) Representation of surface roughness in fused deposition modelling. *Journal of Materials Processing Technology*, 209(15–16):5593–5600.
10. Galantucci LM, Lavecchia F, Percoco G (2009) Experimental study aiming to enhance the surface finish of fused deposition modeled parts. *CIRP Annals*, 58(1):189–192.
11. Sood AK, Ohdar RK, Mahapatra SS (2012) Experimental investigation and empirical modelling of FDM process for compressive strength improvement. *Journal of Advanced Research*, 3(1):81–90.
12. Singh S, Prakash C, Antil P, Singh R, Królczyk G, Pruncu CI (2019) Dimensionless analysis for investigating the quality characteristics of aluminium matrix composites prepared through fused deposition modelling assisted investment casting. *Materials*, 12(12):1907.

13. Singh S, Singh N, Gupta M, Prakash C, Singh R (2019) Mechanical feasibility of ABS/HIPS-based multi-material structures primed by low-cost polymer printer. *Rapid Prototyping Journal*, 25(1):152–161.

14. Singh S, Singh M, Prakash C, Gupta MK, Mia M, Singh R (2019) Optimization and reliability analysis to improve surface quality and mechanical characteristics of heat-treated fused filament fabricated parts. *The International Journal of Advanced Manufacturing Technology*, 102(5–8):1521–1536.

15. Sahu AK, Mahapatra SS, Chatterjee S (2017) Optimization of electrical discharge coating process using MOORA based firefly algorithm. *ASME 2017 Gas Turbine India Conference* (pp. V002T10A005). American Society of Mechanical Engineers.

16. Sahu AK, Mahapatra SS, Chatterjee S, Thomas J (2018) Optimization of surface roughness by MOORA method in EDM by electrode prepared via selective laser sintering process. *Materials Today: Proceedings*, 5:19019–19026.

17. Khan A, Maity K (2016) Parametric optimization of some non-conventional machining processes using MOORA method. *International Journal of Engineering Research in Africa*, 20:19–40.

6 Effect of Process Parameters on Cutting Forces and Osteonecrosis for Orthopedic Bone Drilling Applications

Atul Babbar, Vivek Jain, and Dheeraj Gupta
Thapar Institute of Engineering and Technology

Chander Prakash and Sunpreet Singh
Lovely Professional University

Ankit Sharma
Chitkara University

CONTENTS

6.1   INTRODUCTION

Bones at first glance look lifeless but are active and living tissues being remodelled constantly. It is a composite tissue having a hard outer layer made of crystalline calcium phosphate (hydroxyapatite) with small amounts of other mineral substances, covering a soft spongy structure made of the protein collagen. The outer layer gives strength, and inner honeycomb-like structure containing matrix gives the flexibility required by the body. With structurally supporting the body, it also protects vital organs. They help with providing an environment for the bone marrow to create blood cells and also act as storage banks of many minerals such as calcium. The bones are stiffer and stronger at higher strain rates.

Macroscopically, bones consist of two types of tissues: cortical and cancellous bones. Porosity is the primary feature to distinguish between these two bone types, with the cancellous bone higher in porosity. Cortical bone is considered to have a relative density greater than 0.7, which reflects the presence of minor amounts of porosity, whereas it is lower than 0.7 in case of cancellous bone. Cortical bone (compact bone) is the outer hard layer which is strong, dense, and durable. It is about 4/5th of adult bone mass. Its function is to give supportive strength to the body and protection to organs, and to reserve and liberate chemicals such as calcium. At micro-level, its structure is complex, which plays a crucial role in its mechanical properties. It is laid down in 5 µm thick layers called lamellae. The collagen fibres run parallel to one another within these layers but with a different orientation compared to other layers. The arrangement of lamellae is in different ways in various parts of the bone. Close to the outer and inner surfaces of bone, there is a circumferential arrangement of lamellae parallel to one another, known as circumferential lamellae. In between, there exist cylindrical-shaped structures, approximately 10 mm long and 0.2 mm in diameter, formed from concentric lamellae, named as osteonal bone, which is aligned with the long axis of the bone. A highly unified network of canals and channels is suffused in this solid bone matrix. These networks contain osteocytes, where extracellular fluid supplies dissolved nutrients to them. Concentric lamellae surround *Haversian canal*, through which a blood vessel runs. Then, there exist microscopic channels called *canaliculi*, which help to connect the central canal to these lamellae.

The group comprising a canal, surrounding lamellae, and intervening bone is termed as an osteon. The packed osteons become a proper bone having in-between space called *interstitial lamellar bone*. There is a junction between interstitial bone and osteons named as *cement line*, which is less than 1 µm in thickness. It is a layer which is highly mineralized and collagen-free. Due to the above feature, cortical bone has greater stiffness along with the strength in the horizontal axis of the cortical bone rather than perpendicular direction. Also, it is stronger in compression than in tension [12]. Cancellous bone (trabecular bone) is a network of rod-like structures, which is lighter and more flexible than compact bone. The vertebra, as well as the ends of the long bones, does contain femur, tibia, and radius. It has a lamellar structure same as the cortical bone, with lamellae running parallel to the trabeculae.

6.1.1 Types of Bones

Bones are classified based on their shapes, which are given as follows:

1. Long bones: There length-to-width ratio is very high. It consists of mid diaphysis at the centre and epiphysis at both ends of the bone. This bone facilitates in supporting weight and helps in body movements. Femur and tibia are examples of a long bone.
2. Short bones: These are nearly box-shaped with a thin layer of compact bone surrounding a spongy internal structure. There is an absence of medullary cavity with a very low length-to-width ratio. Wrist and ankle bones are examples of short bones.
3. Flat bones: These are curved and thinner in shape. A layer of spongy bone is sandwiched in between two parallel layers of compact bones. Breastbone and most of the skull parts are flat bones.
4. Sesamoid bones: Tendons constitute these bones, such as patella or knee-cap. Their role is to protect tendons from stress and wear.
5. Irregular bones: These are bones with an unusual shape and do not fit into the previous four categories discussed. These are the bones of the spine and pelvis. Their function is to protect organs or tissues.

6.1.2 Bone Cells

The bone cells found in the human beings are osteoblasts, osteoclasts, osteocytes, and lining cells as illustrated in Figure 6.1.

1. Osteoblasts: These are the "builder" of bones. These are mono-nucleate bone-forming cells. They are found closer to the bone surface and helps in the production of protein mixture known as osteoid, which accumulates and gets harden to result in bone formation.
2. Osteoclasts: These are the "destructor" of bones. These are very large multinucleate cells responsible for the resorption of bones. These are the cells with multiple nuclei located on bone surfaces in resorption pits called *Howships lacunae.*

    They demineralize the adjacent bone with acids by dissolving its collagen. Osteoblasts and osteoclasts are crucial parts of the bone remodelling cycle. One makes and other creates, meaning a continuous change in bone structure takes place with the repair of bone cells.
3. Osteocytes and lining cells: These are created from osteoblasts embedded in the bone matrix during their process of secretion, making them inactive while lining cells are found residing on bone tissue's upper layer, connected by small canals that help them in communication [2]. Osteocytes get trapped and surrounded by bone matrix produced by them. Lining cells reside on the surface of the bone when the formation of bone stops.

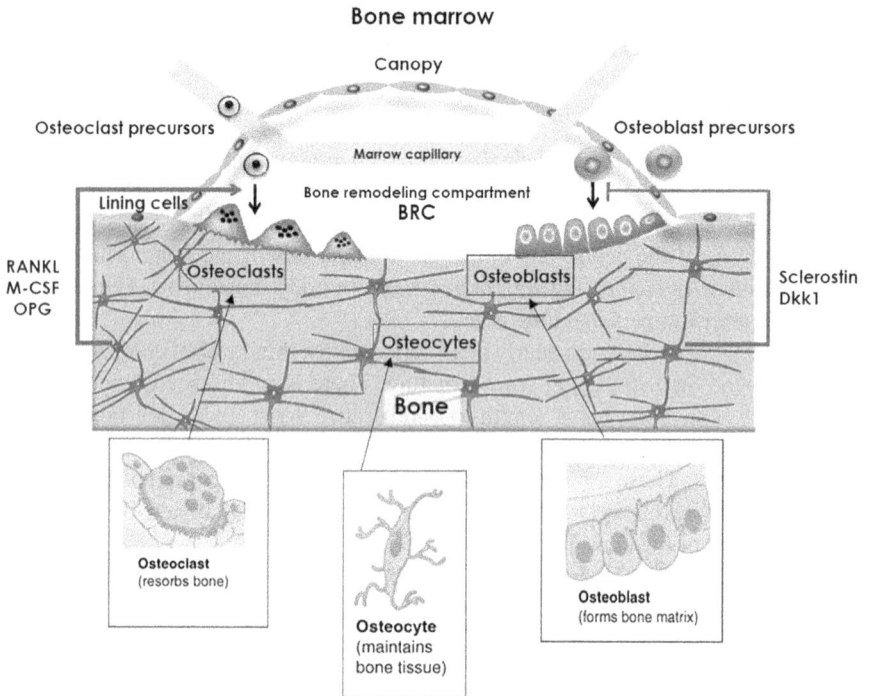

FIGURE 6.1   Bone cells [1].

6.2   BONE DRILLING

A bone fracture occurs due to different reasons such as injury, ageing, or diseases. Bone drilling is a frequently used method for the treatment of bone fractures in which a bone is drilled to create a space for the insertion of screws, plates, and wires, etc. Rehabilitation of various orthopedic surgeries involves restoring the affected bone parts to their position and restraining them until complete healing [3]. For all that, sometimes we need to drill the bones and fasten the screws for easy and quick healing of the bones. This orthopedic drilling is very much similar to mechanical drilling process, which results in the reactive forces and increased temperature of surrounding bone material, which can cause the osteonecrosis in some of the cases and affects the reliability of surgery [4]. A fractured bone is a severe problem faced by a human from the starting of human life on this planet. Self-healing of the bone is a time-consuming process, and sometimes, the bone fixes on the wrong position. So, the allocation of fractured bone at the desired position is a tough task, and in this task, two basic approaches are taken into account: a conventional approach and a direct approach.

In the direct approach, screws are used to fix the damaged bones on their desired position. Before tightening the screws, it requires drilling and tapping of bone [5]. The devices for biomedical applications can be manufactured using different methods

[6–14]. A different actuating mechanism has been used for driving the drill bits such as pneumatic and electric driven tools [15]. In the conventional approach, the fractured bone part is restricted to move relatively from outside support. Traumatologist treats the fracture by fixing it on the desired position and placing the support from outside. With this process, minor cracks and injury can be treated easily and successfully. But in the case of a major dislocation, this process cannot help much better. Different techniques and processes have been used for biomedical applications [16–31].

During this process, heat is generated due to friction between the drill bit and the bone surface. This rise in temperature has been reported to cause the cell's death owing to the carbonization of the cells and leads to the change in physical and chemical properties of the bone [32]. An increase in threshold temperature above 47°C results in the thermal necrosis [33]. This thermal necrosis is just like the death of bone cells and may lead to bone death due to insufficient blood supply [34]. The screw must be properly engaged to grip the bone with a screw. However, the necrotic bone may cause bone's failure and lose the screw, which results in the loss of fixation [35].

6.3 MACHINING PARAMETERS AFFECTING CUTTING FORCES AND TEMPERATURE

Drilling parameters play a very vibrant role for controlling the cutting forces and temperature generated during bone drilling. Parameters associated with the setup of hand drill are the machining parameters and influence the drilling quality and precision. Thermal necrosis also depends upon these machining parameters. The parameters are reported in two major categories:

1. Machining parameters
2. Drill specifications.

6.3.1 MACHINING PARAMETERS

6.3.1.1 Rotational Speed, Feed Rate, and Applied Drill Force

There are numerous factors which contribute to the rise in temperature during drilling of bone such as rotational speed, feed rate, and applied drill force among them. Different researchers have performed many trials to reduce or even eliminate thermal necrosis [36–41]. When one stationary solid and one rotating solid strike with each other, heat is generated owing to the frictional behaviour of solids [42], so optimum rotational speed should be used which results in minimum heat generation during drilling. Researchers suggest a different set of rotational speed for different conditions. The schematic representation of the drilling in cortical bone using tap drill has been shown in Figure 6.2.

Ponnusamy et al. [44] performed conventional drilling on human cadaveric with 3.2 mm diameter drill bit as shown in Figure 6.3. The process parameters used were

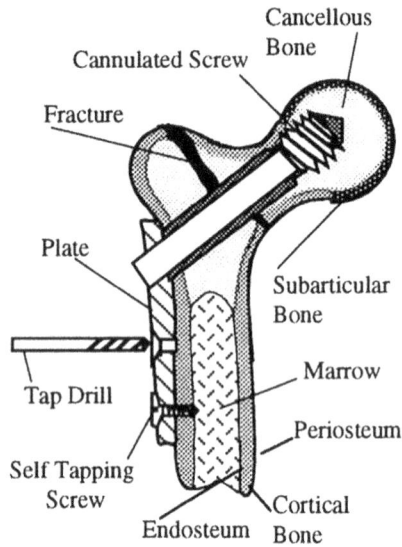

FIGURE 6.2   Drilling in cortical bone with tap drill [43].

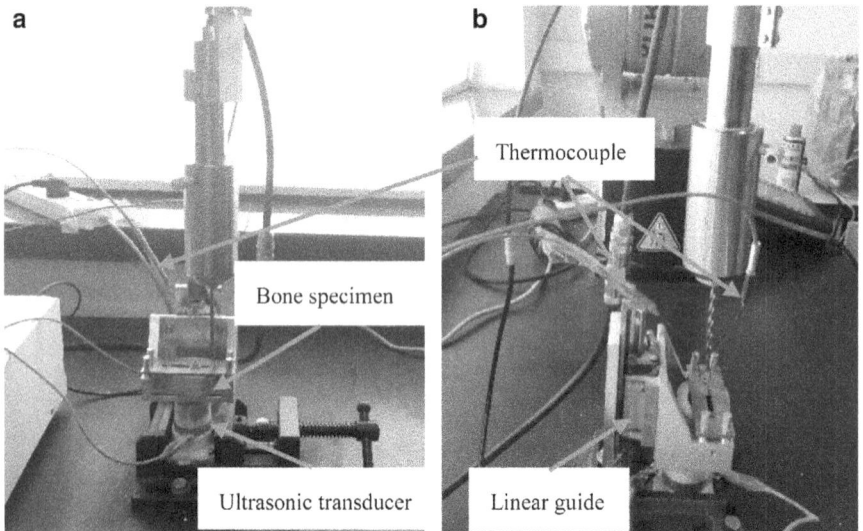

FIGURE 6.3   Experiment setup used for temperature measurement during bone drilling [45].

rotational speed (500, 1000, and 1500 rpm) and feed rate (40, 60, 80 mm/min). The results revealed that maximum temperature increased as the feed rate increased from 40 up to 80 mm/min. But, the in situ temperature decreased when the feed rate increased above 80 mm/min. Temperature-induced during bone drilling is positively correlated with spindle speed. Most significant parameter found to be directly followed by feed rate and then apparent density.

6.3.1.2   Effect of Coolant

Coolant plays an important role in reducing heat generation rate during bone drilling. It has been observed that temperature significantly reduced during drilling [46]. Two methods of drilling are generally used during the drilling process:

1. Internal cooling system
2. External cooling system.

In the internal cooling system, coolant circulates in the tunnel provided in the drill bit as shown in Figure 6.4. When the coolant flows inside the drill bit, the heat generated during the bone drilling gets transferred to the coolant due to the conduction mode of heat transfer [2,34,47], which removes the heat from the machining zone through the removal of bone chips and heat from the drilling zone [48–50]. On another side, coolant is poured in the drilling zone from the outside by sideways as shown in Figure 6.5. One of the concerns related to the use of coolant is that it wets the machining zone due to which bone chips produced during the drilling process get settled over the surface of the tool and cause a significant rise in the temperature due to the increased friction [41,51,52].

6.3.1.3   Depth of Drill

Depth of drill is also a major factor which is to be taken into account before starting the bone drilling. Heat generation during drilling is a key issue which causes some major problems in the recovery of fracture. Presently, depth is estimated by the skilled operator, but if it goes into more depth as compared to required, it will take more post-operative time to recover than the normal. Depth of drill also depends

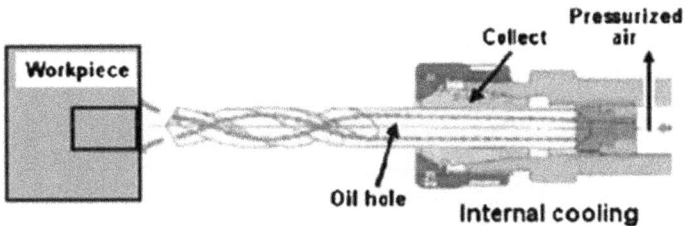

FIGURE 6.4   Systematic arrangement of an internal cooling system [53].

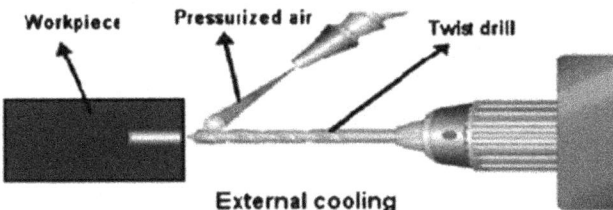

FIGURE 6.5   View of external cooling [53].

on the thickness of bone [54,55]. The mean cortical thickness of the bovine bone is 7–9 mm and human cadaveric bone 3–5 mm [56,57]. Depth of drill is also varied with the density of bone. So, there is a large variation in temperature as we go in depth of the bone.

6.3.1.4 Predrilling and Step Drilling

In predrilling, a hole having a diameter less than the screw is created. Predrilling is always recommended for the hard materials so as to avoid splitting of the work-piece. The diameter of the drill bit is increased from the lower to required diameter in the steps, also known as multistep drilling. This type of drilling is also known as incremental drilling. A single drill bit of required diameter is used in the case of single drilling [58–60]. The drill bit used in step drilling is similar to that used in general drilling. In predrilling, the total time of drilling is increased as compared to step drilling so this type of tool is preferred over predrilling.

6.3.2 DRILL BIT SPECIFICATION

Past researchers have reported that multiple drill bit specifications have a significant influence on thermal injury during bone drilling [36,53,61,62],e.g., the diameter of the drill, helix and clearance angles, drill point, and rake angle. The schematic representation of the drill bit specifications is shown in Figure 6.6.

6.3.2.1 Drill Diameter

Maximum output diameter required after drilling is the major factor, on which other parameters is to be adjusted. Generally, 2.5, 3, and 4 mm diameter drill bits are used to drill the bone [47,54,64,65]. Drill diameter is to be selected according to bone conditions such as density and position of the bone. The diameter of drill also affects the temperature rise during drilling. As the diameter increases, contact surface also increases which increases the friction and hence temperature [54,63,66].

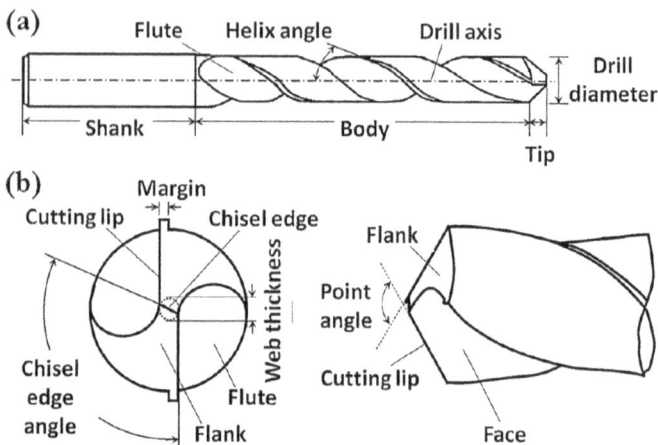

FIGURE 6.6   (a) Twist drill bit and (b) drill bit tip [63].

Furthermore, reduction in the drill diameter may cause breakage or bend of the drill bit.

6.3.2.2   Flutes and Helix Angle

The twist deep groove has been provided around the drill bit referred to as a flute. This provides an easy path for the evacuation of waste material from the drilling zone. However, material removal becomes difficult if these grooves are not provided over the drill bit. These flutes have been provided with varying size and helix angle.

6.3.2.3   Drill Wear

When two hard surfaces slide with each other, some part of the material from the surfaces is removed in the form of small tiny particles. In the case of drilling, there is wear out of cutting edges due to mechanical and thermal effects [4,67–69]. This wear of cutting lips of the drill bit may lead to the increase in axial thrust force, temperature, and vibrations.

6.3.2.4   Cutting Edge Angles

The front edge that majorly involves in cutting of a standard drill bit in any type of drilling includes in cutting face of the drill bit. It includes

1. Rake angle
2. Clearance angle and flank.
1. Rake angle: It is the angle between the cutting edge and the plane perpendicular to the workpiece. It has been observed that the rake angle critically influences the cutting forces [56,70–72]. Past studies recommended an optimum rake of 20°–30°. It has been observed that minimum forces either thrust or tangential have been reported during bone drilling. Furthermore, this recommended angle assists in the easy evacuations of the bone chips amid bone drilling process in various orthopedic operations.
2. Flank and clearance angle: The flat part of the drill bit is known to be a flank part. Flank often leads to the generation of the heat during drilling owing to the friction between the large surface area of the flank and work-piece. However, the undesirable contact of the tool with the workpiece has been prevented by providing clearance angle on the drill bit. But, the clearance provided on the drill bit is often not sufficient to overcome the influence of the large flank surface area in rising the temperature and friction during drilling [73–75].

6.3.2.5   Drill Point

Drill point includes two major parts which helps in efficient drilling and also influences the amount of heat generation. These two parts include

1. Point angle
2. Chisel edge.

Point angle: It is the angle formed in between the projections of cutting edges on the plane passing through the longitudinal axis of the drill bit. It guides the drill bit to an appropriate point where drilling is to be performed. In the case of bone drilling, high precision and accuracy are required to avoid any unwanted circumstances. The smaller point angle provides more acute tip which can easily stab in the bone [55,66,76], whereas more acute tip involves less portion of cutting lip during the first few revolutions of the drill bit and results in higher heat generation rate followed by a rise in temperature. On the other side, when a large point angle is to be used, full contact of cutting lip is involved in the cutting action [53,64,77]. Past researchers have recommended a point angle of 90° for the bone drilling operation.

Chisel edge: The end edge part of the drill bit which assists in asserting the pressure over the workpiece during drilling is known as chisel edge. This is located in the middle of the drill bit. It joins the two cutting edges of the drill bit [52,75]. Chisel edge has a direct relation with thrust force produced at the time of the drill.

6.4 CASE STUDIES OF ROTARY ULTRASONIC DRILLING

Rotary ultrasonic drilling (RUD) is a state-of-art drilling process which provides better control along with reduced cutting forces and temperature, hence eliminating the chances of the osteonecrosis during bone drilling. It removes the material from the workpiece with the combined action of conventional drilling and ultrasonic-assisted drilling process.

Alam et al. [78] experimentally investigated the force and torque during Conventional drilling (CD) and Ultrasonic assisted drilling (UAD) of cortical bone as shown in Figure 6.7. Investigations were carried out by considering four input parameters: rotational speed, feed rate, amplitude, and frequency. It was observed that force decreases by 27% in the range of 600–3000 rpm for CD and by 55% in UAD. There was a significant reduction in the force on decreasing the amplitude, whereas feed rate had less effect on output in UAD. The frequency was varied in the range 10–30 kHz, and the corresponding decrease in force and torque was 57% and 28%, respectively. For 5–15 μm variation in amplitude, the force decreases by 46%, and torque decrease by 14%.

Gupta et al. [79] used rotary ultrasonic machining (RUM) process for drilling in a porcine bone. Investigations were carried out with four input parameters: rotational

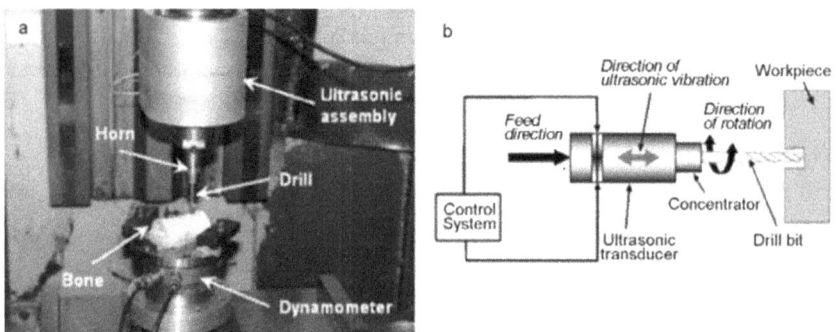

FIGURE 6.7 Rotary ultrasonic drilling setup along with dynamometer for measuring the cutting forces (under Creative Commons CC-BY license) [78].

speed, feed rate, drill diameter, and amplitude. The dynamometer was used to measure the force and torque during drilling. Statistical models were developed, average maximum and minimum forces came out to be 22.9 and 10.08 N, respectively, whereas average maximum and minimum torques were 5.937 and 2.501 N mm. it was concluded that RUM exerts less force and torque in comparison with conventional drilling and produces almost no cracks on the machined surface. It was seen that force and torque decrease with an increase in speed and amplitude while they decrease with a decrease in feed and diameter. The rotational speed came out to be the most influential parameter among all the input parameters which highly affects the output characteristic.

Gupta et al. [80] made a comparative analysis between rotary ultrasonic drilling of bone (RUDB) and conventional surgical bone drilling (CSBD) using hollow diamond-coated abrasive tool and conventional drill bit. Drilling was performed on the porcine bone. Thermocouples were used for temperature measurement, and RUDB showed 40.2°C±0.4°C and 40.3°C±0.2°C, while CSDB showed 74.9°C±0.8°C and 74.9°C±0.6°C. It was concluded that RUDB generates damage-free surface, less force and temperature in comparison with CSDB.

Shakouri et al. [81] compared conventional drilling with the rotary ultrasonic drilling of the bovine femur bone. Investigations were carried out to prevent thermal necrosis during drilling of bone with input variables cutting speed (500, 750, 1000, 1500, 2000) and feed rate (500, 100, 150), while the depth of cut, frequency, and amplitude were kept constant. The results demonstrated that rotary ultrasonic drilling significantly reduces the temperature and cutting force in comparison with conventional drilling. It was seen that temperature rise in vibration-assisted drilling was more at a low rotational speed. But while increasing the rotational speed, the rise in temperature decreases and the thermal injury was minimum at a rotational speed of 1000 rpm and a feed rate of 100 and 150 mm/min. The feed rate had no significant effect on the rise in the temperature.

Li et al. [82] carried out ultrasonic-assisted drilling on the cat skull bone having a dimension of 10 mm×4 mm×1.5 mm with a tool diameter of 300 μm, a rotational speed of 14 m/s, and a feed rate of 0.2 mm/s. The ultrasonic frequency used was 29.7 kHz with a maximum vibrational amplitude of 75 μm. It has been found that ultrasonic vibration with higher amplitude caused no damage to bone dura mater interface and no leakage of CSF liquid. Moreover, no chips generated and there is a reduced risk of infection on the actuating tool with ultrasonic vibrations.

6.5   CONCLUSION

Bone drilling is generally related to the rise in temperature which may cause osteonecrosis when the temperature rises beyond the physiological levels. This rise in temperature can initiate thermogenesis and can initiate various sorts of thermal injuries to the hard and soft tissues. It has been observed that osteonecrosis depends upon the elevated temperature and duration of the exposure to that temperature. Owing to the lower thermal conductivity of the bone, the heat generated during drilling causes a severe rise in the temperature at the drilling site. Therefore, a controlled set of optimized process parameters should be used to reduce cutting forces and thermal necrosis. All efforts should be made to maintain the temperature below the consensus threshold temperature of 47°C during bone drilling, above which osteonecrosis reported. Osteoclastic resorption can cause severe implications during implant fixation. The dulling and wear of the cutting flutes owing to the repetitive use of the drill bits caused a rise in the cutting forces, and drill bit can break instead of advancing into the bone. Ultrasonic-assisted drilling has been proved to be beneficial in terms of better control, reduced cutting forces, and minimizing the rise in temperature during in vivo drilling. Further developments are required to benefit the patients during in vivo surgical operations.

REFERENCES

1. T. Bellido, L.I. Plotkin and A. Bruzzaniti, Basic and applied bone biology, in *Basic and Applied Bone Biology*, 2014. Academic Press. https://doi.org/10.1016/C2011-0-05817-9
2. S. Harder, C. Egert, S. Freitag-Wolf, C. Mehl and M. Kern, Intraosseous temperature changes during implant site preparation: in vitro comparison of thermocouples and infrared thermography, *Int. J. Oral Maxillofac. Implants* 33 (2018), pp. 72–78.
3. S. Takenaka, N. Hosono, Y. Mukai, K. Tateishi and T. Fuji, Significant reduction in the incidence of C5 palsy after cervical laminoplasty using chilled irrigation water, *Bone Joint J.* 98B (2016), pp. 117–124.
4. C.J. Chauhan, D.N. Shah and F.B. Sutaria, Various bio-mechanical factors affecting heat generation during osteotomy preparation: a systematic review, *Indian J. Dent. Res.* 29 (2018), pp. 81–92.
5. K. Tamai, A. Suzuki, S. Takahashi, J. Akhgar, M.S. Rahmani, K. Hayashi et al., The incidence of nerve root injury by highspeed drill can be reduced by chilled saline irrigation in a rabbit model, *Bone Joint J.* 99B (2017), pp. 554–560.
6. A. Sharma, A. Babbar, V. Jain and D. Gupta, Enhancement of surface roughness for brittle material during rotary ultrasonic machining, in *MATEC Web of Conferences*, Vol. 249, 2018, p. 01006. EDP Sciences.
7. A. Babbar, V. Jain and D. Gupta, Thermogenesis mitigation using ultrasonic actuation during bone grinding: a hybrid approach using CEM43°C and Arrhenius model, *J. Brazilian Soc. Mech. Sci. Eng.* 41 (2019), pp. 401.
8. M. Kumar, A. Babbar, A. Sharma and A.S. Shahi, Effect of post weld thermal aging (PWTA) sensitization on micro-hardness and corrosion behavior of AISI 304 weld joints, *J. Phys. Conf. Ser.* 1240 (2019), p. 012078.
9. D. Singh, A. Babbar, V. Jain, D. Gupta, S. Saxena and V. Dwibedi, Synthesis, characterization, and bioactivity investigation of biomimetic biodegradable PLA scaffold fabricated by fused filament fabrication process, *J. Brazilian Soc. Mech. Sci. Eng.* 41 (2019), pp. 121.
10. A. Babbar, P. Singh and H.S. Farwaha, Parametric study of magnetic abrasive finishing of UNS C26000 flat brass plate, *Int. J. Adv. Mechatronics Robot.* 9 (2017), pp. 83–89.
11. A. Babbar, V. Jain and D. Gupta, Neurosurgical bone grinding, in *Biomanufacturing*, Springer International Publishing, Cham, 2019, pp. 137–155. doi:https://doi.org/10.1007/978-3-030-13951-3_7
12. A. Babbar, A. Sharma, V. Jain and A.K. Jain, Rotary ultrasonic milling of C/SiC composites fabricated using chemical vapor infiltration and needling technique, *Mater. Res. Express* 6 (2019), p. 085607.
13. R. Baraiya, A. Babbar, V. Jain and D. Gupta, In-situ simultaneous surface finishing using abrasive flow machining via novel fixture, *J. Manuf. Process.* 50 (2020), pp. 266–278.

14. A. Babbar, A. Kumar, V. Jain and D. Gupta, Enhancement of activated tungsten inert gas (A-TIG) welding using multi-component TiO2-SiO2-Al2O3 hybrid flux, *Measurement* 148 (2019), p. 106912.
15. S. Olson, J.M. Clinton, Z. Working, J.R. Lynch, W.J. Warme, W. Womack et al., Thermal effects of glenoid reaming during shoulder arthroplasty in vivo, *J. Bone Jt. Surg. – Ser. A* 93 (2011), pp. 11–19.
16. A.A. Aliyu, A.M. Abdul-Rani, T.L. Ginta, C. Prakash, E. Axinte, M.A. Razak et al., A review of additive mixed-electric discharge machining: current status and future perspectives for surface modification of biomedical implants, *Adv. Mater. Sci. Eng.* (2017), pp. 1–23.
17. C. Prakash, S. Singh, B.S. Pabla, S.S. Sidhu and M.S. Uddin, Bio-inspired low elastic biodegradable Mg-Zn-Mn-Si-HA alloy fabricated by spark plasma sintering, *Mater. Manuf. Process.* 34 (2019), pp. 357–368.
18. C. Prakash, H.K. Kansal, B.S. Pabla, S. Puri, C. Prakash, H.K. Kansal et al., Experimental investigations in powder mixed electric discharge machining of Ti – 35Nb – 7Ta – 5Zr β – titanium alloy, *Mater. Manuf. Process.* 32 (2017), pp. 274–285.
19. C. Prakash and M.S. Uddin, Surface modification of β-phase Ti implant by hydroaxyapatite mixed electric discharge machining to enhance the corrosion resistance and in-vitro bioactivity, *Surf. Coatings Technol.* 326 (2017), pp. 134–145.
20. C. Prakash, S. Singh, K. Verma, S.S. Sidhu and S. Singh, Synthesis and characterization of Mg-Zn-Mn-HA composite by spark plasma sintering process for orthopedic applications, *Vacuum* 155 (2018), pp. 578–584.
21. S. Singh, C. Prakash and S. Ramakrishna, 3D printing of polyether-ether-ketone for biomedical applications, *Eur. Polym. J.* 114 (2019), pp. 234–248.
22. C. Prakash, S. Singh, M. Singh, M.K. Gupta, M. Mia and A. Dhanda, Multi-objective parametric appraisal of pulsed current gas tungsten arc welding process by using hybrid optimization algorithms, *Int. J. Adv. Manuf. Technol.* 101 (2019), pp. 1107–1123.
23. C. Prakash, H.K. Kansal, B.S. Pabla, S. Puri and A. Aggarwal, Electric discharge machining – a potential choice for surface modification of metallic implants for orthopedic applications: a review, *Proc. Inst. Mech. Eng. Part B J. Eng. Manuf.* 230 (2016), pp. 331–353.
24. C. Prakash, S. Singh, B.S. Pabla and M.S. Uddin, Synthesis, characterization, corrosion and bioactivity investigation of nano-HA coating deposited on biodegradable Mg-Zn-Mn alloy, *Surf. Coatings Technol.* 346 (2018), pp. 9–18.
25. S. Singh, M. Singh, C. Prakash, M.K. Gupta, M. Mia and R. Singh, Optimization and reliability analysis to improve surface quality and mechanical characteristics of heat-treated fused filament fabricated parts, *Int. J. Adv. Manuf. Technol.* 102 (2019), pp. 1521–1536.
26. C. Prakash, H.K. Kansal, B.S. Pabla and S. Puri, Processing and characterization of novel biomimetic nanoporous bioceramic surface on β-Ti implant by powder mixed electric discharge machining, *J. Mater. Eng. Perform.* 24 (2015), pp. 3622–3633.
27. C. Prakash, S. Singh, R. Singh, S. Ramakrishna and S. Puri, *Biomanufacturing*, Springer International Publishing, Cham, 2019.
28. Poomathi N, Singh S, Prakash C, Patil RV, Perumal PT, Barathi VA, Balasubramanian KK, Ramakrishna S, Maheshwari NU., et al., Bioprinting in ophthalmology: current advances and future pathways. *Rapid Prototyp. J.* (2019).
29. C. Prakash, H.K. Kansal, B.S. Pabla and S. Puri, On the Influence of nanoporous layer fabricated by PMEDM on β - Ti implant : biological and computational evaluation of bone- implant interface, *ScienceDirect Mater. Today Proc.* 4 (2017), pp. 2298–2307.
30. S. Singh, N. Singh, M. Gupta, C. Prakash and R. Singh, Mechanical feasibility of ABS/HIPS-based multi-material structures primed by low-cost polymer printer, *Rapid Prototyp. J.* 25 (2019), pp. 152–161.

31. C. Prakash, S. Singh, M. Gupta, M. Mia, G. Królczyk and N. Khanna, Synthesis, characterization, corrosion resistance and in-vitro bioactivity behavior of biodegradable Mg–Zn–Mn–(Si–HA) composite for orthopaedic applications, *Materials* 11 (2018), p. 1602.

32. S.K. Chauhan and V.R. Singh, Loss of strength in drilled bone in orthopaedic surgery, *Biomed. Mater. Eng.* 1 (1991), pp. 251–253.

33. H. Heydari, N. Cheraghi Kazerooni, M. Zolfaghari, M. Ghoreishi and V. Tahmasbi, Analytical and experimental study of effective parameters on process temperature during cortical bone drilling, *Proc. Inst. Mech. Eng. Part H J. Eng. Med.* 232 (2018), pp. 871–883.

34. V. Gupta, P.M. Pandey, R.K. Gupta and A.R. Mridha, Rotary ultrasonic drilling on bone: a novel technique to put an end to thermal injury to bone, *Proc. Inst. Mech. Eng. Part H J. Eng. Med.* 231 (2017), pp. 189–196.

35. K. Alam, I.M. Bahadur and N. Ahmed, Cortical bone drilling: an experimental and numerical study, *Technol. Health Care* 23 (2014), pp. 223–231.

36. R.K. Pandey and S.S. Panda, Evaluation of delamination in drilling of bone, *Med. Eng. Phys.* 37 (2015), pp. 657–664.

37. K.N. Bachus, M.T. Rondina and D.T. Hutchinson, The effects of drilling force on cortical temperatures and their duration: an in vitro study, *Med. Eng. Phys.* 22 (2000), pp. 685–691.

38. J. Sui, N. Sugita, K. Ishii, K. Harada and M. Mitsuishi, Mechanistic modeling of bone-drilling process with experimental validation, *J. Mater. Process. Tech.* 214 (2014), pp. 1018–1026.

39. L. Lamazza, G. Garreffa, D. Laurito, M. Lollobrigida, L. Palmieri and A. De Biase, Temperature values variability in piezoelectric implant site preparation: differences between cortical and corticocancellous bovine bone, *Biomed Res. Int.* 2016 (2016), pp. 1–7.

40. R.K. Pandey and S.S. Panda, Optimization of bone drilling parameters using grey-based fuzzy algorithm, *Measurement* 47 (2014), pp. 386–392.

41. A. Cseke and R. Heinemann, The effects of cutting parameters on cutting forces and heat generation when drilling animal bone and biomechanical test materials, *Med. Eng. Phys.* 51 (2018), pp. 24–30.

42. R.K. Pandey and S.S. Panda, Optimization of multiple quality characteristics in bone drilling using grey relational analysis, *J. Orthop.* 12 (2015), pp. 39–45.

43. M.T. Hillery and I. Shuaib, Temperature effects in the drilling of human and bovine bone, *J. Mater. Process. Technol.* 92–93 (1999), pp. 302–308.

44. P. Pandithevan, N.V.M. Pandy and C. Palanivel, Development of in-situ temperature prediction models from cadaveric human femur for bone drilling, *J. Mech. Med. Biol.* 18 (2018), p. 1850026.

45. Z. Sun, Y. Wang, K. Xu, G. Zhou, C. Liang and J. Qu, Experimental investigations of drilling temperature of high-energy ultrasonically assisted bone drilling, *Med. Eng. Phys.* 65 (2019), pp. 1–7.

46. N. Bertollo and W. Robert, Drilling of bone: practicality, limitations and complications associated with surgical drill-bits, in *Biomechanics in Applications*, InTech, 2011, pp. 53-83.

47. K. Gok, L. Buluc, U.S. Muezzinoglu and Y. Kisioglu, Development of a new driller system to prevent the osteonecrosis in orthopedic surgery applications, *J. Brazilian Soc. Mech. Sci. Eng.* 37 (2015), pp. 549–558.

48. E. Bagci and B. Ozcelik, Effects of different cooling conditions on twist drill temperature, *Int. J. Adv. Manuf. Technol.* 34 (2007), pp. 867–877.

49. P. Trisi, M. Berardini, A. Falco, M. Podaliri Vulpiani and G. Perfetti, Insufficient irrigation induces peri-implant bone resorption: an in vivo histologic analysis in sheep, *Clin. Oral Implants Res.* 25 (2014), pp. 696–701.

50. E. Shakouri, H. Haghighi Hassanalideh and S. Gholampour, Experimental investigation of temperature rise in bone drilling with cooling: a comparison between modes of without cooling, internal gas cooling, and external liquid cooling, *Proc. Inst. Mech. Eng. Part H J. Eng. Med.* 232 (2018), pp. 45–53.

51. Z. Zhi-Jin, J.-Z. Zhu, B.J. Schaller, R. Gruber, H.A. Merten, T. Kruschat et al., Thermal Analysis of Grinding, *Proc. Inst. Mech. Eng. Part B J. Eng. Manuf.* 1 (2012), pp. 101–107.

52. A. Feldmann, J. Wandel and P. Zysset, Reducing temperature elevation of robotic bone drilling, *Med. Eng. Phys.* 38 (2016), pp. 1495–1504.

53. R. Kumar and S.S. Panda, Drilling of bone : a comprehensive review, *J. Clin. Orthop. Trauma* 4 (2013), pp. 15–30.

54. G. Augustin, T. Zigman, S. Davila, T. Udilljak, T. Staroveski, D. Brezak et al., Cortical bone drilling and thermal osteonecrosis, *Clin. Biomech.* 27 (2012), pp. 313–325.

55. C. Yeager, A. Nazari and D. Arola, Machining of cortical bone: surface texture, surface integrity and cutting forces, Mach. *Sci. Technol.* 12 (2008), pp. 100–118.

56. A.T. Berman, J.S. Reid, D.R. Yanicko, G.C. Sih and M.R. Zimmerman, Thermally induced bone necrosis in rabbits. Relation to implant failure in humans, *Clin. Orthop. Relat. Res.* (1984), pp. 284–92.

57. Y.C. Chen, Y.K. Tu, Y.J. Tsai, Y.S. Tsai, C.Y. Yen, S.C. Yang et al., Assessment of thermal necrosis risk regions for different bone qualities as a function of drilling parameters, *Comput. Methods Programs Biomed.* 162 (2018), pp. 253–261.

58. Y.X. Yang, C.Y. Wang, Z. Qin, L.L. Xu, Y.X. Song and H.Y. Chen, Drilling force and temperature of bone by surgical drill, *Adv. Mater. Res.* 126 (2010), pp. 779–784.

59. G. Singh, V. Jain and D. Gupta, Comparative study for surface topography of bone drilling using conventional drilling and loose abrasive machining, *Proc. Inst. Mech. Eng. Part H J. Eng. Med.* 229 (2015), pp. 225–231.

60. M.B. Abouzgia and D.F. James, Temperature rise during drilling through bone, *Int. J. Oral Maxillofac. Implants* 12 (1997), pp. 342–53.

61. R.K. Pandey and S.S. Panda, Multi-performance optimization of bone drilling using Taguchi method based on membership function, *Measurement* 59 (2015), pp. 9–13.

62. R.K. Pandey and S.S. Panda, Bone drilling: an area seeking for improvement, in *2011 Nirma University International Conference on Engineering: Current Trends in Technology*, NUiCONE 2011- Conference Proceedings, 2011, pp. 8–10.

63. J.E. Lee, Y. Rabin and O.B. Ozdoganlar, A new thermal model for bone drilling with applications to orthopaedic surgery, *Med. Eng. Phys.* 33 (2011), pp. 1234–1244.

64. T. MacAvelia, M. Salahi, M. Olsen, M. Crookshank, E.H. Schemitsch, A. Ghasempoor et al., Biomechanical measurements of surgical drilling force and torque in human versus artificial femurs, *J. Biomech. Eng.* 134 (2012), p. 124503.

65. G. Augustin, S. Davila, T. Udiljak, D.S. Vedrina and D. Bagatin, Determination of spatial distribution of increase in bone temperature during drilling by infrared thermography: preliminary report, *Arch. Orthop. Trauma Surg.* 129 (2009), pp. 703–709.

66. V. Gupta, P.M. Pandey, A.R. Mridha and R.K. Gupta, Effect of various parameters on the temperature distribution in conventional and diamond coated hollow tool bone drilling: a comparative study, *Procedia Eng.* 184 (2017), pp. 90–98.

67. S.C. Mohlhenrich, A. Modabber, T. Steiner, D.A. Mitchell and F. Holzle, Heat generation and drill wear during dental implant site preparation: systematic review, *Br. J. Oral Maxillofac. Surg.* 53 (2015), pp. 679–689.

68. V. Gupta and P.M. Pandey, In-situ tool wear monitoring and its effects on the performance of porcine cortical bone drilling: a comparative in-vitro investigation, *Mech. Adv. Mater. Mod. Process.* 3 (2017), pp. 2.

69. T. Staroveski, D. Brezak and T. Udiljak, Drill wear monitoring in cortical bone drilling, *Med. Eng. Phys.* 37 (2015), pp. 560–566.

70. A. Feldmann, P. Ganser, L. Nolte and P. Zysset, Orthogonal cutting of cortical bone: temperature elevation and fracture toughness, *Int. J. Mach. Tools Manuf.* 118–119 (2017), pp. 1–11.

71. N. Yusup, A.M. Zain and S.Z.M. Hashim, Evolutionary techniques in optimizing machining parameters: review and recent applications (2007–2011), *Expert Syst. Appl.* 39 (2012), pp. 9909–9927.

72. C. Ma, E. Shamoto, T. Moriwaki and L. Wang, Study of machining accuracy in ultrasonic elliptical vibration cutting, *Int. J. Mach. Tools Manuf.* 44 (2004), pp. 1305–1310.

73. N. Senthilkumar, T. Tamizharasan and V. Anandakrishnan, Experimental investigation and performance analysis of cemented carbide inserts of different geometries using Taguchi based grey relational analysis, *Measurement* 58 (2014), pp. 520–536.

74. M.K. Gupta, G. Singh and P.K. Sood, Modelling and optimization of tool wear in machining of EN24 steel using Taguchi approach, *J. Inst. Eng. Ser. C* 96 (2015), pp. 269–277.

75. A. Feldmann, M. Schweizer, S. Stucki and L. Nolte, Experimental evaluation of cortical bone substitute materials for tool development, surgical training and drill bit wear investigations, *Med. Eng. Phys.* 66 (2019), pp. 107–112.

76. G. Augustin, S. Davila, K. Mihoci, T. Udiljak, D.S. Vedrina and A. Antabak, Thermal osteonecrosis and bone drilling parameters revisited, *Arch. Orthop. Trauma Surg.* 128 (2008), pp. 71–77.

77. G.J.M. Tuijthof, C. Frühwirt and C. Kment, Influence of tool geometry on drilling performance of cortical and trabecular bone, *Med. Eng. Phys.* 35 (2013), pp. 1165–1172.

78. K. Alam, A.V. Mitrofanov and V.V. Silberschmidt, Experimental investigations of forces and torque in conventional and ultrasonically-assisted drilling of cortical bone, *Med. Eng. Phys.* 33 (2011), pp. 234–239.

79. V. Gupta and P.M. Pandey, An in-vitro study of cutting force and torque during rotary ultrasonic bone drilling, *Proc. Inst. Mech. Eng. Part B J. Eng. Manuf.* 232 (2018), pp. 1549–1560.

80. V. Gupta and P.M. Pandey, Experimental investigation and statistical modeling of temperature rise in rotary ultrasonic bone drilling, *Med. Eng. Phys.* 38 (2016), pp. 1330–1338.

81. E. Shakouri, M.H. Sadeghi, M.R. Karafi, M. Maerefat and M. Farzin, An in vitro study of thermal necrosis in ultrasonic-assisted drilling of bone, *Proc. Inst. Mech. Eng. Part H J. Eng. Med.* 229 (2015), pp. 137–149.

82. M. Yang, C. Li, Y. Zhang, Y. Wang, B. Li and Y. Hou, Experimental research on microscale grinding temperature under different nanoparticle jet minimum quantity cooling, *Mater. Manuf. Process.* 32 (2017), pp. 589–597.

7 Fabrication and Machining Methods of Composites for Aerospace Applications

Atul Babbar
Shree Guru Gobind Singh Tricentenary University, Gurugram, Haryana, India

Vivek Jain and Dheeraj Gupta
Thapar Institute of Engineering and Technology

Chander Prakash
Lovely Professional University

Ankit Sharma
Chitkara University

CONTENTS

7.1 INTRODUCTION

The engineering industry has been constantly evolving over the years through several innovations and inventions. The increase in the number of complex engineering applications has led to the development of many advanced materials such as advanced ceramics and high strength temperature-resistant (HSTR) alloys. [1]. The modern engineering industries are constantly dealing with these advanced materials to transform them to suit the various applications. The process of transforming these materials involves various forms of material removal from them. The material removal from these advanced materials is a great challenge faced by the modern engineering industry.

Among the advanced materials, ceramics have extensive applications due to their superior wear resistance and refractoriness [2]. The most important characteristic of ceramics that possess difficulty in machining is the brittleness or the fragile nature of the ceramics which causes them to crack easily while machining. Furthermore, they are also prone to thermal cracking at high temperatures that arise during the machining process at the tool–work interface due to the high-temperature gradient that develops in their structure [3]. Machining of material can be generally categorized into two categories: conventional (traditional) and non-conventional machining. Conventional machining contains direct contact of workpiece and tool [4]. However, highly brittle and hard material is difficult to machine by conventional machining techniques such as turning, drilling, and shaping [5]. Nonconventional machining processes have surpassed the limitations associated with conventional machining processes. In recent years, rotary ultrasonic milling has emerged as a technique which can machine the material with superior machining characteristics. It provides decreased cutting force, lesser generation of heat, and increased tool life [6].

7.2 COMPOSITES

Composites are defined as the material in which two or more constituents have been brought together to produce a material with entirely different properties from its parent components. The nature and properties of the composites are determined by the interface between the components whereas the strength of the composites is generally governed by the adhesion force which can be either chemical, physical, or combination of both [7]. Different types of processes have been implemented till now in manufacturing applications [8–26].

7.2.1 Types of Composites

7.2.1.1 Metal Matrix Composite (MMC)

MMC is the material which is formed by the combination of a metal with a metal or any other material. Strength and wear of MMC material can be enhanced by the reinforcement of the metal matrix [27]. Generally, a matrix is of lighter metal such as aluminium, titanium, and magnesium.

7.2.1.2 Polymer Matrix Composite (PMC)

PMC is the combination of polymer matrix with fibrous reinforcement dispersed phase [28]. The density of PMCs is less than the metal and ceramics. They are corrosion resistive in nature and have high insulation.

7.2.1.3 Ceramic Matrix Composites (CMCs)

CMC is the combination of ceramic fibre with the ceramic matrix. Fibre-reinforced ceramic matrix composites have high toughness and high strength [29]. CMCs are formed by the combination of a ceramic fibre which is implant into a ceramic matrix to generate a finished CMC part. The C/SiC composite possesses excellent wear, oxidation, and thermal shock resistance [30]. It has superior properties such as lightweight, high specific strength, low density, and abrasion resistance [31]. Due to their good mechanical properties, it has been used as a high-temperature structural material in an aeroplane engine, heat exchangers, racing cars, etc. [32]. It was used by NASA as a flat component on X Series space machines [29,33] and also used in the high-temperature oven.

In the modern world, the efficiency of cars, turbine engines, and the airplane is attributed on the development of ceramic-based chamber which can combat much higher temperature (approx. 1500°C) as compared to their metallic counterparts (1000°C). Indian Space Research Organization (ISRO) has been using C/SiC for making nose cap, wing leading edge ,and control surfaces of aerospace vehicle RLV-TD (reusable launch vehicle—Technology Demonstrator) where temperature encountered much higher than others like silica tile.

Another area where high-temperature capability of CMCs is useful is the thrust chamber of satellite thrusters. In satellite thrusters, high-temperature alloys have been used [34]. The permissible operating temperature is 1300°C, whereas, in C/SiC ceramic thrusters with suitable functionally gradient coating, the operating temperature is up to 1900°C which will enhance the impulse of the thruster. Germany and China have already developed C/SiC nozzle extension and satellite thrust chambers.

7.3 METHODS OF PREPARATION OF CERAMIC MATRIX COMPOSITE

7.3.1 Polymer Infiltration and Pyrolysis (PIP)

In polymer infiltration and pyrolysis (PIP), initially pyrolysing of fibre preform is carried out at a temperature of above 900°C in an inert gas atmosphere. This pyrolysing converts the preform into the porous preform. After this, capillary force was used to infiltrate the molten matrix material in a vacuum atmosphere at a temperature of more than 1450°C. But composites formed by PIP have weak fibre matrix interfacial bonding. So, the porosity of composite is quite high.

7.3.2 REACTION BONDING

In the reaction bonding technique, a multi-layer of two different materials was formed by adding thin layers via a chemical reaction. Subsequently, the multi-layer material is placed inside the layers of the substrate at low temperature. Since the temperature is low, the degradation of fibre is very low. In this process, the densification was avoided, so the possibility of shrinkage is reduced.

7.3.3 HOT PRESSING

In hot pressing, powder or compacted material is placed inside the die, and at high temperature (more than 2000°C), uniform pressure was applied across the die. This will make the material hard. The main limitation of hot pressing is the shape of the material. As die is required, it is only suitable for simple shapes. Making a new die for each shape is not commercially viable.

7.3.4 CHEMICAL VAPOUR INFILTRATION

In chemical vapour infiltration (CVI), fibre-reinforced composite is made by infiltrating matrix material at high temperature in reactive gases environment into the fibrous preform. CVI is a similar technique like chemical vapour deposition (CVD). The CVD is used for the deposition process on the hot bulk surface, whereas CVI is for deposition on the porous surface. In CVD, the deposition of material is done by pyrolysing the gaseous pressure, on the external surface of a bulk substrate. CVI is also CVD but on the internal surface of a substrate. CVD is commonly used for augment substrate surface and to deposit conformal films where conventional techniques for surface modification are not competent. CVI is highly useful in atomic layer deposition process for depositing enormously thin layers of material. A CVI process differs from physical vapour deposition (PVD) processes, such as reactive sputtering and evaporation. PVD involves adsorption of molecular and atomic species on the substrate. The schematic representation of the CVI system has been shown in Figure 7.1.

In CVI process, the chemical reaction of the precursor gases takes place both on the substrate and in the gas phase. Reactions can be infiltrated by higher frequency radiations like as UV (photo-assisted CVI), plasma (plasma-enhanced CVI), or through heat (thermal CVI). The CVI process is highly complex in nature and includes different gas phases and surface reactions. This process contains a "boundary layer," i.e. a hot layer of precursor gases just above the substrate. The gas-phase pyrolysis reactions which occur inside the boundary layer play an important part in the deposition process. In thermal CVI process, the growth rate of the film is mainly dependent on the operating pressure of the CVI reactor, chemical composition and interaction of gas-phase, and the temperature of the substrate. In CVI system, first, transportation and evaporation of precursors occur in the bulk gas flow area inside the reactor. Gaseous by-products and reactive intermediates are generated by the gas-phase reactions of precursors inside the reaction zone of the reactor followed by the transportation of reactants on the surface of the substrate. Adhesion of the atoms

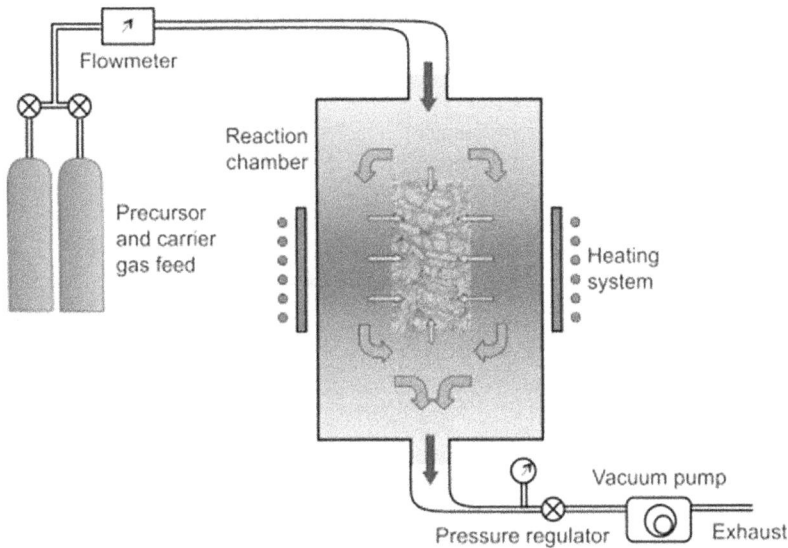

FIGURE 7.1   Schematic representation of CVI system [35].

or molecules (i.e. adsorption) of the gaseous reactants is on to the surface of the substrate. Subsequently, surface diffusion and nucleation occur which leads to the formation of the film. At last, desorption and transportation of waste material take place, i.e. remains of the decomposition, outside the reaction zone.

7.4   PROCESSING OF FIBRE-REINFORCED CMC VIA CVI PROCESS

The CVI process is very useful in the fabrication of fibre-reinforced CMCs such as C/SiC, C/C-SiC, and SiC/SIC composites. In the CVI process, a fibrous preform is prepared and is placed inside the reactor of CVI. This preform works as the reinforcement material. The stages involved in the CVI process have been demonstrated in the Figure 7.2. The vapours or reactant gases were supplied inside the reactor that diffused into the fibrous preform. The illustration for CVI growth has been shown in Figure 7.3. A composite material formed by the decomposition of these reactants, as they covered the void space between the fibres [36]. In the composite, deposited material works as the matrix and fibrous preform act as reinforcement. The diameter of fibre increases with the progress in reactions.

7.4.1   PROCESSING OF CMCs

Fibre-reinforced CMCs are being fabricated by producing a dense ceramic matrix around carbon or silicon carbide filters. The fabrication of these CMCs is mainly done by different techniques using gaseous reactants (CVI) or by using polymer reactants, i.e. PIP, or by using molten elements reacting with the preforms (reactive melt infiltration (RMI) or liquid silicon infiltration (LSI)), or by using high-pressure and high-temperature sintering of ceramic (HP-Sinter, high-pressure sinter process) [37–39]. CVI process has many distinct advantages for

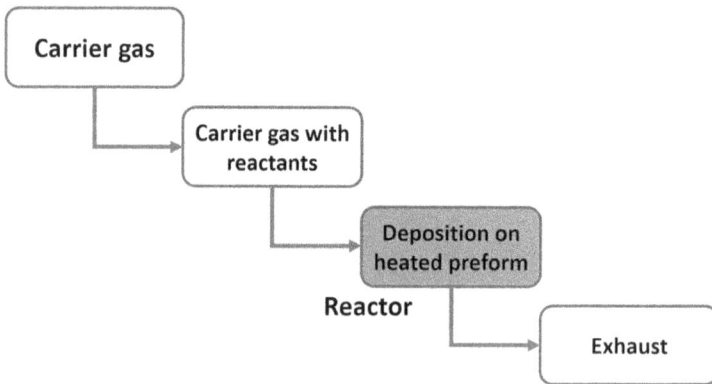

FIGURE 7.2   Stages in CVI process.

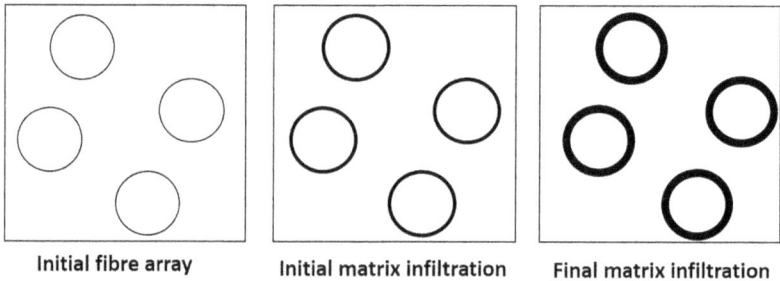

| Initial fibre array | Initial matrix infiltration | Final matrix infiltration |

FIGURE 7.3   CVI growth.

the fabrication of CMCs such as uniform coating, less mechanical damage, purer surface, high thermal, and chemical resistance [40]. Materials densified by CVI do not require post-treatment to eliminate the organics (as required in the polymer infiltration process). Moreover, due to the presence of low infiltration temperature, CVI produces low residual stress [41].

7.4.1.1 CVI Reactor

CVI reactor works as the core body in the CVI process. The schematic diagram of the reactor is shown in Figure 7.4. CVI reactor contains three important parts: (i) a feed system, (ii) an effluent system, and (iii) a heating chamber.

7.4.2 Types of CVI Processes

7.4.2.1 Isothermal-Isobaric Infiltration

It is the oldest hot wall technique which is widely used in industries as well as laboratories. In isothermal infiltration, gases are entered by the diffusion. These gases surround the whole porous substrate (fibre). The generated material has low residual stress and contains a uniform microstructure. Its main drawback is its slow processing time, i.e. sometimes more than 500 h.

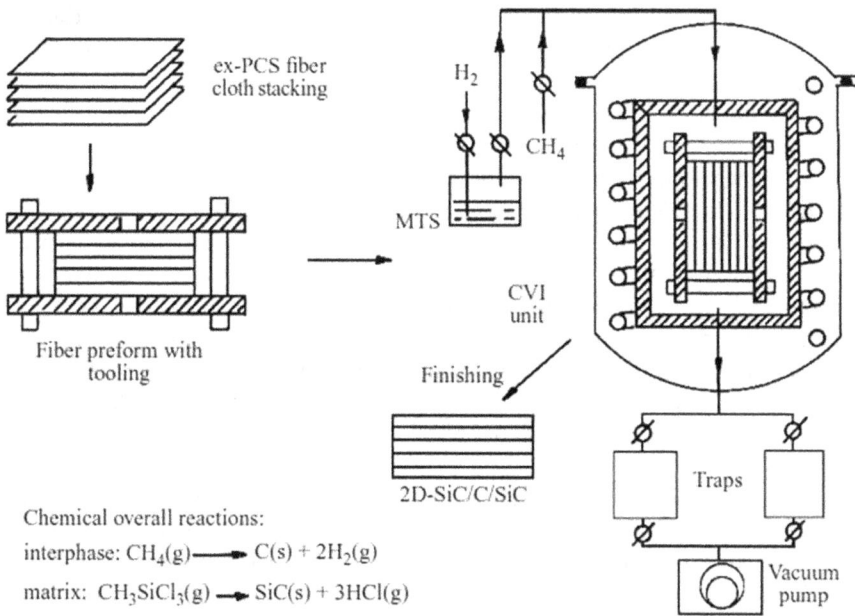

ex-PCS fiber
cloth stacking

Fiber preform with
tooling

H_2

CH_4

MTS

CVI
unit

Finishing

2D-SiC/C/SiC

Traps

Vacuum
pump

Chemical overall reactions:

interphase: $CH_4(g) \longrightarrow C(s) + 2H_2(g)$

matrix: $CH_3SiCl_3(g) \longrightarrow SiC(s) + 3HCl(g)$

FIGURE 7.4 CVI reactor and process [42].

7.4.2.2 Thermal Gradient

In this process, heat is supplied only on one side of the fibrous structure. Reactant gases entered from the cold side, and deposition will happen on the hotter side. This creates a mobile reacting front where the decomposition of vapour occurs that produce the deposit.

7.4.2.3 Pressure Gradient

In pressure gradient CVI, the precursor gases are forced through the porous preform by applying a pressure gradient across the preform. This process can generate a ceramic matrix composite in very lesser time.

7.4.2.4 Film Boiling

In this process, the preform is dipped into a liquid precursor. The liquid precursor is recycled back after condensation due to which the efficiency of deposition is high at the cost of high- power requirements.

7.5 ROTARY ULTRASONIC MACHINING (RUM)

7.5.1 BACKGROUND OF RUM

In the present scenario, continuous development and research have contributed in the development of new tailor-made and hard materials [43]. The machining of such materials is quite cumbersome. Another group of materials, such as silicon, ferrites, glass, quartz, germanium, sapphire, ceramics, corundum, and some composites, are not easily machinable because of greater hardness and brittleness [44–46]. Ceramic matrix composites are also hard to machine composite material such as C/SiC [47]. The requirements for machining of these advanced materials have led to arise of non-conventional machining techniques [31–60]. Past researchers have implemented different techniques for the removing material from the workpiece [36,48–55]. Ultrasonic machining (USM) is one of the non-traditional machining processes which is also known as subtraction manufacturing process [56].

Figure 7.5 is a schematic illustration of the USM process. The tool oscillates at a high frequency (>20 kHz) and is continuously fed towards the workpiece with high pressure [57]. The abrasive slurry is supplied between the tip of tool and workpiece which consist of tiny abrasive particles and water. Material removal occurs in the form of small particles due to the continuous impacting of the abrasive particles onto the workpiece [58]. However, one of the major limitations of USM is low material removal rate and less dimensional accuracy [59].

This limitation has been surpassed by the hybrid RUM process which combines ultrasonic machine and diamond grinding for combined action towards material removal. USM was patented in 1927 and has been continuously used in the industry since 1940 for machining materials with high brittleness and hardness [61]. This process uses abrasive slurry, a mixture of diamond abrasives, and a cooling fluid that is fed between an ultrasonically vibrating tool and workpiece [62]. RUM was developed as an improvement over USM. Unlike USM, instead of using the loose abrasive slurry, the diamond abrasives were impregnated on the rotating tool. P. Legge developed RUM for

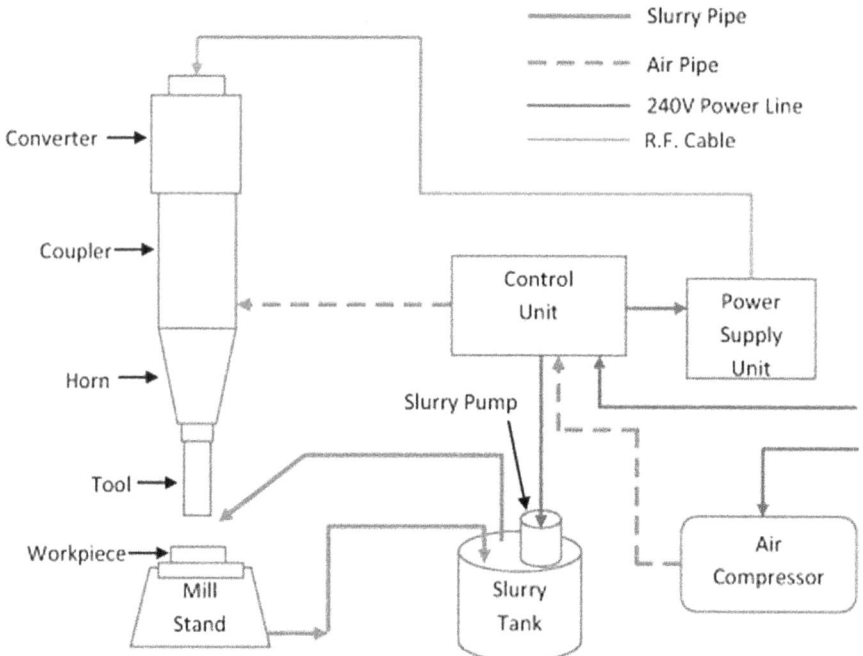

FIGURE 7.5   Illustration of the ultrasonic machine (under a creative common distribution) [60].

the first time in 1964 [63]. Continuous research also increased the application of RUM. Face milling of ceramics can also be done with RUM [64], and subsequently, higher subtraction of the material from workpiece has been noticed than the USM as well as diamond grinding [65,66]. RUM uses a metal-bonded tool for subtraction of the material from the workpiece instead of abrasive slurry using combined action of rotation and ultrasonic actuation with applied frequency and amplitude [67]. The tool continuously approaches towards the stationary workpiece with provided feed rate. A metal-bonded tool (diamond coated abrasive tool) which is rotating as well as ultrasonically vibrating is fed into the workpiece with either constant pressure or feed rate [68]. The coolant used during the machining has numerous important functions such as cooling the drill bits, removing the debris from the machining zone, preventing drill bit jamming, and making operation smooth [69]. The RUM process is illustrated in Figure 7.6.

Experimental results have shown that under similar condition, RUM gives six to ten times higher machining rate than conventional grinding process [6]. It is ten times faster than the USM, and RUM can drill deep and small holes easily compared to USM [71]. It also provides low tool pressure and improved hole accuracy [72].

7.6 CASE STUDIES

Ding et al. [33] performed rotary USM of C/SiC composites. Furthermore, these results were compared with conventional drilling (CD). The authors found that RUM can produce high-quality exit hole at a higher feed rate, but CD can give good hole quality at a

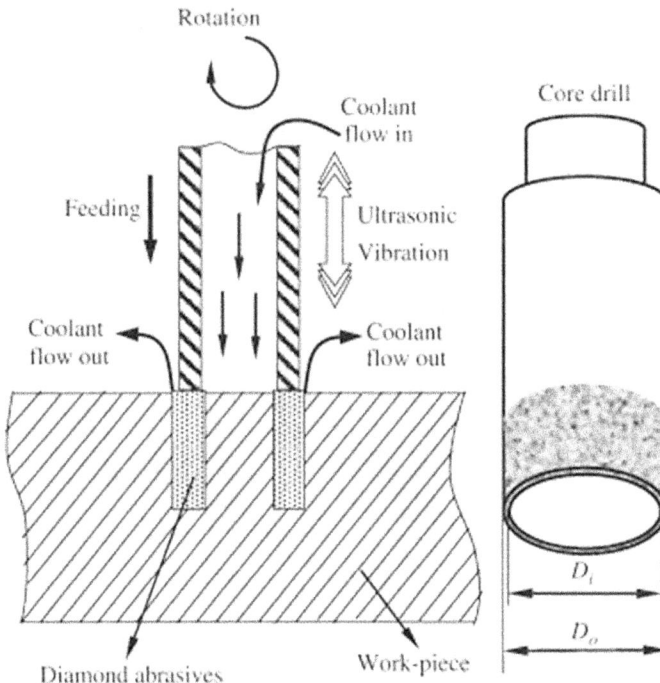

FIGURE 7.6 Illustration of the rotary ultrasonic machine [70].

low feed rate and specific high spindle. Drilling force and torque used by RUM decrease by 23% and 47.6%, respectively, as shown in Figures 7.7 and 7.8. It was observed that RUM produces better machining quality since the degree of tearing of hole determined by drilling force, so hole exit is better at low drilling force. Furthermore, it was reported that surface roughness reduces 23% in case of RUM compared to CD. This is because of the reduction in the size of the chip. This reduction in chip size also suggests less brittleness of C/SiC. The reduction percentage of torque and drilling force increased progressively with the spindle speed, while the change with the feed rate is very minor.

Kadivar et al. [73] combined ultrasonic vibration (22 kHz frequency) and CD during the machining of Al/SiC composites. The composites were prepared using powder extrusion technique. Different process variables such as rotational speed (125, 710, and 1400 rev/min) and feed rate (0.08, 0.18, and 0.32 rev/min) have been

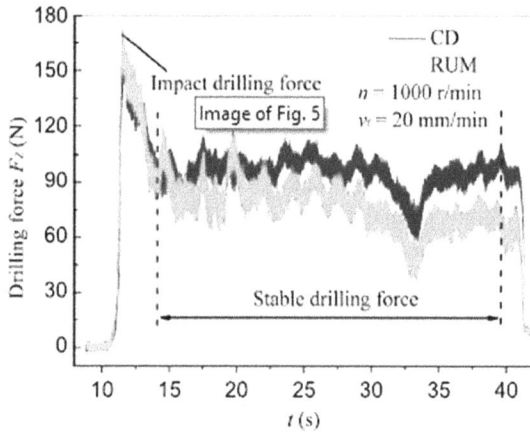

FIGURE 7.7 Comparison of drilling force between CD and RUM [33].

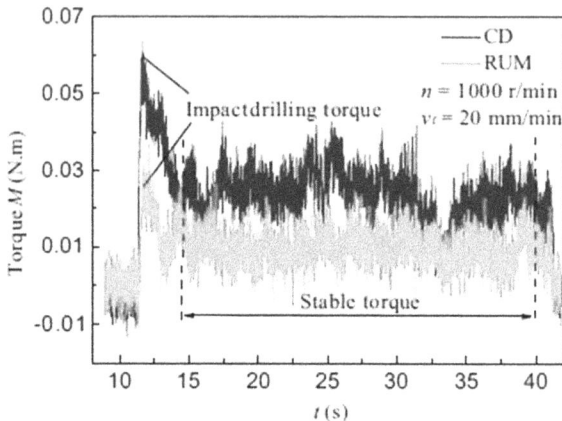

FIGURE 7.8 Compassion of torque between CD and RUM [33].

investigated to reduce the burr size. It was found that during CD, burr height showed increasing trend with the increase in the feed rate. Furthermore, smaller burr height was achieved on applying ultrasonic vibration to the drilling tool. Overall, 83% and 24% reduction in the burr height has been reported in RUM compared to CD, and this highlights the potential of the RUM process.

Babbar et al. [36] used RUM process for the milling of C/SiC composites. First, composites were fabricated by CVI and needling process followed by milling using RUM process. It was found that maximum change in surface roughness is 6.08 µm and MRR is 66.82%. The speed, feed rate, and depth of cut were had significant effect on the response characteristics.

Abdo et al. [74] optimized RUM parameters during face milling of zirconia ceramics. The influence of amplitude, frequency, spindle speed, depth of cut, and feed rate on the surface roughness was studied. A spindle speed of 2000, 4000, and 6000 rpm, a feed rate of 50, 100, and 150 mm/min, a vibration frequency of 20, 21.5, 23 and kHz, a power supply of 50%, 60%, and 70%, and a depth of cut of 0.025, 0.05, and 0.075 mm were used. The authors have used a diamond milling cutter of an outer diameter of 4 mm and an inner diameter 1.5 mm. It was noticed that as the rotational speed is increased or feed rate decreased, the surface roughness showed decreasing trend. Moreover, surface roughness increases slightly on increasing depth of cut. The optimized machining parameters for minimum Ra value of 0.4295 µm found to be 6000 rpm rotational speed, 50 mm/min feed rate, 23 KHz frequency, 60% power, and 0.025 mm depth of cut.

V. Garcia Navas [75] investigated RUM during machining of Al_2O_3-TiC-SiC ceramics, which was fabricated by the spark plasma sintering method (SPS). A diamond milling tool with varying grain size was used. In face milling operation, maximum material removal was found at 5000 rpm, a feed rate of 150 mm/min, and a_p of 0.2 mm, and in slot milling, maximum material removal without workpiece damage was at 5000 rpm, a feed of 300 mm/min, and a_p of 0.3 mm. In finishing operation during face milling, optimum surface roughness was at 5000 rpm, a feed of 100 mm/min, and a_p of 0.05 mm, whereas in slot milling at 5000 rpm, a feed rate of 50 mm/min, and a_p of 0.05 mm. It was concluded that lower compressive residual stress will be generated on increasing feed and decreasing axial depth of cut. The authors reported that lower Ra value can be achieved with a diamond tool having smaller grain size, higher feed rate, and higher depth of cut.

7.7   CONCLUSION

In this chapter, the key issues governing the fabrication and machining methods of the carbon fibre-reinforced composites have been discussed. The material removal phenomena of the CRFP composites are different from the other conventional metals and alloys. The material behaviour towards the machining depends upon diverse reinforcement, orientation, matrix properties, cutting direction, and relative matrix content. The fabrication of the composites using CVI is a promising method which allows fabricating the ceramic composites with uniform grain distribution, little damage to the fibre structure, reduced stresses, and high mechanical, thermal, and corrosion resistance. The machining methods using the state-of-art designed cutting tools are the need of the present scenario. The optimal cutting parameters can help in substantially improving the tool life. It has been observed that higher cutting speed leads to the reduced cutting forces and hence has a prolonged effect on the useful life of the tool. The extensive research is under investigation for machining the CFRP composites for aerospace applications using promising high-speed milling, orbital drilling process, minimum quantity lubrication, and machining using specialized diamond grinding tools. In all these aspects, rotary ultrasonic milling has given remarkable results as far as the machining of ceramic composites is concerned. The RUM not only reduces cutting forces in comparison with the conventional machining methods but also results in decreasing the thermal and mechanical defects.

REFERENCES

1. E. Bertsche, K. Ehmann and K. Malukhin, An analytical model of rotary ultrasonic milling, *Int. J. Adv. Manuf. Technol.* 65 (2013), pp. 1705–1720.
2. N.T. Nguyen, Aerospace applications, in *Advanced Textbooks in Control and Signal Processing*, 2018, pp. 349–429.
3. Y. Wang, H. Guo and S. Gong, Thermal shock resistance and mechanical properties of $La_2Ce_2O_7$ thermal barrier coatings with segmented structure, *Ceram. Int.* 35 (2009), pp. 2639–2644.
4. R. Teti, Machining of composite materials, *CIRP Ann.* 51 (2002), pp. 611–634.
5. Z.Y. Wang, K.P. Rajurkar, J. Fan, S. Lei, Y.C. Shin and G. Petrescu, Hybrid machining of Inconel 718, *Int. J. Mach. Tools Manuf.* 43 (2003), pp. 1391–1396.
6. N.J. Churi, Z.J. Pei, D.C. Shorter and C. Treadwell, Rotary ultrasonic machining of silicon carbide: designed experiments, *Int. J. Manuf. Technol. Manag.* 12 (2007), p. 284.
7. J.G. Taylor, Composites, in *Phenolic Resins: A Century of Progress*, Springer, Berlin, 2010, pp. 263–306. https://link.springer.com/book/10.1007/978-3-642-04714-5
8. S. Chander, S.K. Mishra, P. Chauhan and Ajai, Ice height and backscattering coefficient variability over greenland ice sheets using SARAL radar altimeter, *Mar. Geod.* 38 (2015), pp. 466–476.
9. S. Singh, C. Prakash, M. Singh, G.S. Mann, M.K. Gupta, R. Singh Ramakrishna S et al., Poly-lactic-acid: potential material for bio-printing applications, in *Biomanufacturing*, Springer, Cham, 2019, pp. 69–87.
10. C. Prakash, S. Singh, C.I. Pruncu, V. Mishra, G. Królczyk, D.Y. Pimenov et al., Surface modification of Ti-6Al-4V alloy by electrical discharge coating process using partially sintered Ti-Nb electrode, *Materials* 12 (2019), p. 1006.
11. P.K. Rathod and M. Chander, Concentrate feeding to dairy cattle in India: practices and implications for Indian dairy industry, *Indian J. Anim. Sci.* 86 (2016), pp. 206–212.
12. M.K. Gupta, C.I. Pruncu, M. Mia, G. Singh, S. Singh, C. Prakash et al., Machinability investigations of Inconel-800 super alloy under sustainable cooling conditions, *Materials* 11 (2018), p. 2088.
13. C. Prakash, S. Singh, M. Singh, P. Antil, A.A.A. Aliyu, A.M. Abdul-Rani et al., Multi-objective optimization of MWCNT mixed electric discharge machining of Al–30SiC p MMC using particle swarm optimization, in *Futuristic Composites*, 2018, pp. 145–164.
14. C. Prakash, S. Singh, S. Ramakrishna, G. Królczyk and C.H. Le, Microwave sintering of porous Ti–Nb-HA composite with high strength and enhanced bioactivity for implant applications, *J. Alloys Compd.* 824 (2020), p. 153774.
15. C. Prakash, H.K. Kansal, B.S. Pabla and S. Puri, To optimize the surface roughness and microhardness of β-Ti alloy in PMEDM process using Non-dominated Sorting Genetic Algorithm-II, in *2015 2nd International Conference on Recent Advances in Engineering and Computational Sciences*, RAECS 2015, 2016.

16. C. Prakash, H.K. Kansal, B.S. Pabla and S. Puri, Experimental investigations in powder mixed electric discharge machining of Ti–35Nb–7Ta–5Zrβ-titanium alloy, *Mater. Manuf. Process.* 32 (2017), pp. 274–285.
17. S. Chander, P. Chauhan and Ajai, Variability of altimetric range correction parameters over indian tropical region using JASON-1 & JASON-2 radar altimeters, *J. Indian Soc. Remote Sens.* 40 (2012), pp. 341–356.
18. C. Prakash, H.K. Kansal, B. Pabla, S. Puri and A. Aggarwal, Electric discharge machining – a potential choice for surface modification of metallic implants for orthopedic applications: a review, *Proc. Inst. Mech. Eng. Part B J. Eng. Manuf.* 230 (2016), pp. 331–353.
19. C. Prakash, S. Singh, I. Farina, F. Fraternali and L. Feo, Physical-mechanical characterization of biodegradable Mg-3Si-HA composites, *PSU Res. Rev.* 2 (2018), pp. 152–174.
20. C. Prakash, S. Singh, S. Sharma, J. Singh, G. Singh, M. Mehta et al., Fabrication of low elastic modulus Ti50Nb30HA20 alloy by rapid microwave sintering technique for biomedical applications, *Mater. Today Proc.* 21 (2020), pp. 1713–1716.
21. C. Prakash, S. Singh, A. Basak, G. Królczyk, A. Pramanik, L. Lamberti et al., Processing of Ti50Nb50-xHAx composites by rapid microwave sintering technique for biomedical applications, *J. Mater. Res. Technol.* 9 (2019), pp. 242–252.
22. C. Prakash, H.K. Kansal, B.S. Pabla and S. Puri, Potential of silicon powder-mixed electro spark alloying for surface modification of β-phase titanium alloy for orthopedic applications, *Mater. Today Proc.* 4 (2017), pp. 10080–10083.
23. C. Prakash, S. Singh, S. Sharma, H. Garg, J. Singh, H. Kumar et al., Fabrication of aluminium carbon nano tube silicon carbide particles based hybrid nano-composite by spark plasma sintering, *Mater. Today Proc.* 21 (2019), pp. 1637–1642.
24. C. Prakash, H.K. Kansal, B.S. Pabla and S. Puri, Multi-objective optimization of powder mixed electric discharge machining parameters for fabrication of biocompatible layer on β-Ti alloy using NSGA-II coupled with Taguchi based response surface methodology, *J. Mech. Sci. Technol.* 30 (2016), pp. 4195–4204.
25. C. Prakash, H.K. Kansal, B.S. Pabla and S. Puri, On the influence of nanoporous layer fabricated by PMEDM on β-Ti implant: biological and computational evaluation of bone-implant interface, *Mater. Today Proc.*, 4 (2017), pp. 2298–2307.
26. S. Singh, C. Prakash and M.K. Gupta, On friction-stir welding of 3D printed thermoplastics, in *Materials Forming, Machining and Post Processing*, Springer, Cham, 2020, pp. 75–91.
27. C.T. Lynch and J.P. Kershaw, *Metal Matrix Composites*, CRC Press, USA, 2018.
28. M. Balasubramanian, Polymer matrix composites, in *Composite Materials and Processing*, CRC Press, USA, 2013, pp. 167–266. https://doi.org/10.1201/b15551
29. H. Hocheng, N.H. Tai and C.S. Liu, Assessment of ultrasonic drilling of C/SiC composite material, *Compos. Part A Appl. Sci. Manuf.* 31 (2000), pp. 133–142.
30. K. Zhao, K. Li and Y. Wang, Rapid densification of C/SiC composite by incorporating SiC nanowires, *Compos. Part B Eng.* 45 (2013), pp. 1583–1586.
31. E.J. Lee, D.H. Lee, J.C. Kim and D.J. Kim, Densification behavior of high purity SiC by hot pressing, *Ceram. Int.* 40 (2014), pp. 16389–16392.
32. J.-C. Bae, K.-Y. Cho, D.-H. Yoon, S.-S. Baek, J.-K. Park, J.-I. Kim et al., Highly efficient densification of carbon fiber-reinforced SiC-matrix composites by melting infiltration and pyrolysis using polycarbosilane, *Ceram. Int.* 39 (2013), pp. 5623–5629.
33. K. Ding, Y. Fu, H. Su, Y. Chen, X. Yu and G. Ding, Experimental studies on drilling tool load and machining quality of C/SiC composites in rotary ultrasonic machining, *J. Mater. Process. Technol.* 214 (2014), pp. 2900–2907.
34. D.J. Anderson, E. Pencil, D. Vento, T. Peterson, J. Dankanich, D. Hahne et al., Products from NASA's in-space propulsion technology program applicable to low-cost planetary missions, *Acta Astronaut.* 93 (2014), pp. 516–523.

35. G.L. Vignoles, Chemical vapor deposition/ infiltration processes for ceramic composites, in *Advances in Composites Manufacturing and Process Design*, Woodhead Publishing, UK, 2015, pp. 147–176.
36. A. Babbar, A. Sharma, V. Jain and A.K. Jain, Rotary ultrasonic milling of C/SiC composites fabricated using chemical vapor infiltration and needling technique, *Mater. Res. Express* 6 (2019), p. 085607.
37. P. Delhaes, Chemical vapor deposition and infiltration processes of carbon materials, *Carbon* 40 (2002), pp. 641–657.
38. M. Kütemeyer, L. Schomer, T. Helmreich, S. Rosiwal and D. Koch, Fabrication of ultra high temperature ceramic matrix composites using a reactive melt infiltration process, *J. Eur. Ceram. Soc.* 36 (2016), pp. 3647–3655.
39. P. Sangsuwan, J.A. Orejas, J.E. Gatica, S.N. Tewari and M. Singh, Reaction-bonded silicon carbide by reactive infiltration, *Ind. Eng. Chem. Res.* 40 (2001), pp. 5191–5198.
40. L. Zhang, A.J. Patil, L. Li, A. Schierhorn, S. Mann, U. Gösele et al., Chemical infiltration during atomic layer deposition: metalation of porphyrins as model substrates, *Angew. Chemie – Int. Ed.* 48 (2009), pp. 4982–4985.
41. H. Wang, X. Zhou, J. Yu, Y. Cao and R. Liu, Fabrication of SiC$_f$/SiC composites by chemical vapor infiltration and vapor silicon infiltration, *Mater. Lett.* 64 (2010), pp. 1691–1693.
42. R.R. Naslain, Ceramic matrix composites: matrices and processing, in *Encyclopedia of Materials: Science and Technology*, Elsevier, UK, 2001, pp. 1060–1066.
43. J.P. Davim, *Machining of Hard Materials*, Springer, London, 2011.
44. V. Kumar and H. Singh, Rotary ultrasonic drilling of silica glass BK-7: microstructural investigation and process optimization through TOPSIS, *Silicon* 11 (2019), pp. 471–485.
45. V. Kumar and H. Singh, Investigation of hole quality in rotary ultrasonic drilling of borosilicate glass using RSM, *J. Brazilian Soc. Mech. Sci. Eng.* 41 (2019), p. 36.
46. Y. Hu and H. Wang, Surface grinding of optical BK7/K9 glass using rotary ultrasonic machining: an experimental study, *ASME 2017 12th International Manufacturing Science and Engineering Conference collocated with the JSME/ASME 2017 6th International Conference on Materials and Processing*, 1 (2017), pp. 1–7.
47. I.M. Low, *Advances in Ceramic Matrix Composites*, Woodhead Publishing, 2014.
48. R. Baraiya, A. Babbar, V. Jain and D. Gupta, In-situ simultaneous surface finishing using abrasive flow machining via novel fixture, *J. Manuf. Process.* 50 (2020), pp. 266–278.
49. D. Singh, A. Babbar, V. Jain, D. Gupta, S. Saxena and V. Dwibedi, Synthesis, characterization, and bioactivity investigation of biomimetic biodegradable PLA scaffold fabricated by fused filament fabrication process, *J. Brazilian Soc. Mech. Sci. Eng.* 41 (2019), p. 121.
50. A. Babbar, A. Kumar, V. Jain and D. Gupta, Enhancement of activated tungsten inert gas (A-TIG) welding using multi-component TiO2-SiO2-Al2O3 hybrid flux, *Measurement* 148 (2019), p. 106912.
51. M. Kumar, A. Babbar, A. Sharma and A.S. Shahi, Effect of post weld thermal aging (PWTA) sensitization on micro-hardness and corrosion behavior of AISI 304 weld joints, *J. Phys. Conf. Ser.* 1240 (2019), p. 012078.
52. A. Babbar, V. Jain and D. Gupta, Neurosurgical bone grinding, in *Biomanufacturing*, Springer International Publishing, Cham, 2019, pp. 137–155.
53. A. Sharma, A. Babbar, V. Jain and D. Gupta, Enhancement of surface roughness for brittle material during rotary ultrasonic machining, *MATEC Web Conf.* 249 (2018), p. 01006.

54. A. Babbar, P. Singh and H.S. Farwaha, Parametric study of magnetic abrasive finishing of UNS C26000 flat brass plate, *Int. J. Adv. Mechatronics Robot.* 9 (2017), pp. 83–89.

55. A. Babbar, V. Jain and D. Gupta, Thermogenesis mitigation using ultrasonic actuation during bone grinding: a hybrid approach using CEM43°C and arrhenius model, *J. Brazilian Soc. Mech. Sci. Eng.* 41 (2019), p. 401.

56. J. Kumar, J.S. Khamba and S.K. Mohapatra, An investigation into the machining characteristics of titanium using ultrasonic machining, *Int. J. Mach. Mach. Mater.* 3 (2008), p. 143.

57. J. Kumar, Ultrasonic machining–a comprehensive review, *Mach. Sci. Technol.* 17 (2013), pp. 325–379.

58. T.B. Thoe, D.K. Aspinwall and M.L.H. Wise, Review on ultrasonic machining, *Int. J. Mach. Tools Manuf.* 38 (1998), pp. 239–255.

59. Y.C. Lin, B.H. Yan and Y.S. Chang, Machining characteristics of titanium alloy (Ti–6Al–4V) using a combination process of EDM with USM, *J. Mater. Process. Technol.* 104 (2000), pp. 171–177.

60. D. Goswami and S. Chakraborty, Parametric optimization of ultrasonic machining process using gravitational search and fireworks algorithms, *Ain Shams Eng. J.* 6 (2015), pp. 315–331.

61. C. Nath and M. Rahman, Effect of machining parameters in ultrasonic vibration cutting, *Int. J. Mach. Tools Manuf.* 48 (2008), pp. 965–974.

62. C. Ma, E. Shamoto, T. Moriwaki and L. Wang, Study of machining accuracy in ultrasonic elliptical vibration cutting, *Int. J. Mach. Tools Manuf.* 44 (2004), pp. 1305–1310.

63. W.L. Cong, Z.J. Pei, X. Sun and C.L. Zhang, Rotary ultrasonic machining of CFRP: a mechanistic predictive model for cutting force, *Ultrasonics* 54 (2014), pp. 663–675.

64. Y. Jiao, W.J. Liu, Z.J. Pei, X.J. Xin and C. Treadwell, Study on edge chipping in Rotary ultrasonic machining of ceramics: an integration of designed experiments and finite element method analysis, *J. Manuf. Sci. Eng. Trans. ASME* 127 (2005), pp. 752–758.

65. P. Fernando, M. Zhang, Z. Pei and W. Cong, Intermittent and continuous rotary ultrasonic machining of K9 glass: an experimental investigation, *J. Manuf. Mater. Process.* 1 (2017), p. 20.

66. F.D. Ning, W.L. Cong, Z.J. Pei and C. Treadwell, Rotary ultrasonic machining of CFRP: a comparison with grinding, *Ultrasonics* 66 (2016), pp. 125–132.

67. W.L. Cong, Z.J. Pei, E. Van Vleet, N. Mohanty and C. Treadwell, Vibration amplitude in rotary ultrasonic machining: a novel measurement method and effects of process variables, *J. Manuf. Sci. Eng. Trans. ASME* 133 (2011), pp. 1–5.

68. J. Wang, P. Feng, J. Zheng and J. Zhang, Improving hole exit quality in rotary ultrasonic machining of ceramic matrix composites using a compound step-taper drill, *Ceram. Int.* 42 (2016), pp. 13387–13394.

69. M. Zhou, M. Wang and G. Dong, Experimental investigation on rotary ultrasonic face grinding of SiCp/Al composites, *Mater. Manuf. Process.* 31 (2016), pp. 673–678.

70. D. Liu, W.L. Cong, Z.J. Pei and Y. Tang, A cutting force model for rotary ultrasonic machining of brittle materials, *Int. J. Mach. Tools Manuf.* 52 (2012), pp. 77–84.

71. F. Ning, H. Wang, W. Cong and P.K.S.C. Fernando, A mechanistic ultrasonic vibration amplitude model during rotary ultrasonic machining of CFRP composites, *Ultrasonics* 76 (2017), pp. 44–51.

72. Q. Wang, W. Cong, Z.J. Pei, H. Gao and R. Kang, Rotary ultrasonic machining of potassium dihydrogen phosphate (KDP) crystal: an experimental investigation on surface roughness, *J. Manuf. Process.* 11 (2009), pp. 66–73.

73. M.A. Kadivar, R. Yousefi, J. Akbari, A. Rahi and S.M. Nikouei, Burr size reduction in drilling of Al/SiC metal matrix composite by ultrasonic assistance, *Adv. Mater. Res.* 410 (2011), pp. 279–282.

74. B.M.A. Abdo, S.M. Darwish and A.M. El-Tamimi, Parameters optimization of rotary ultrasonic machining of zirconia ceramic for surface roughness using statistical taguchi's experimental design, *Appl. Mech. Mater.* 184–185 (2012), pp. 11–17.

75. V.G. Navas, A. Sandá, C. Sanz, D. Fernández, J. Vleugels, K. Vanmeensel et al., Surface integrity of rotary ultrasonic machined ZrO_2–TiN and Al2O3–TiC–SiC ceramics, *J. Eur. Ceram. Soc.* 35 (2015), pp. 3927–3941.

8 Computational Analysis of Aerodynamics Characteristics of High-Speed Moving Vehicle

Pawan Singh, Vibhanshu Chhettri, and Nitin Kumar Gupta
DIT University

CONTENTS

8.1 INTRODUCTION

When objects move through air, forces are generated by the relative motion between air and surfaces of the body; the study of these forces generated by the motion of air is called aerodynamics. In our project, various aerodynamic effects due to different changes in the design and the methodology of spoilers are considered. Computational fluid dynamics (CFD) being one of the efficient methodologies to determine the results is given high importance in our study.

An airfoil or aerofoil is the cross-sectional shape of a wing, blade, or sail. When an airfoil-shaped body moved through a fluid process, it produces an aerodynamic force. The component of this force perpendicular to the direction of motion is called lift. The component parallel to the direction of motion is called drag.

Drag Force: Some energy is lost to move the car through the air, and this energy is used to overcome a drag force. In vehicle aerodynamics, drag is due to frontal pressure and rear vacuum. Figure 8.1 shows the schematic diagram representing various forces.

Spoiler: Figure 8.2 shows the spoiler which acts like a barrier to air flow, in order to increase the air pressure in front of the trunk of car. Spoilers are used to produce high pressure that pushes the car down and also give stability at corners.

Diffuser: A diffuser, in an automotive context, is a shaped section of the car under body which improves the car's aerodynamic properties. The primary purpose of a rear diffuser is to efficiently increase the down force of a vehicle. This helps increase grip and reduce aerodynamic drag. Figure 8.3 represents the effect of drag force by diffuser.

Siren: One of the commonly used vehicle add-ons is sirens over police/ambulance cars; Sirens are mounted on fixed locations and used to warn of natural disasters or attacks. They also produce additional aerodynamic drag. They have notable impact on aerodynamic drag depending on cross-wind effects. Figure 8.4 shows the effect of siren on a moving car.

Roof Rack: Figure 8.5 shows that modern passenger cars use various add-ons including roof rack/ski rack/bicycle rack, for commercial and professional reasons. Such add-ons thus cause extra drag and have significant impact on aerodynamic drag.

Lift and Downforce From Over Body Flow

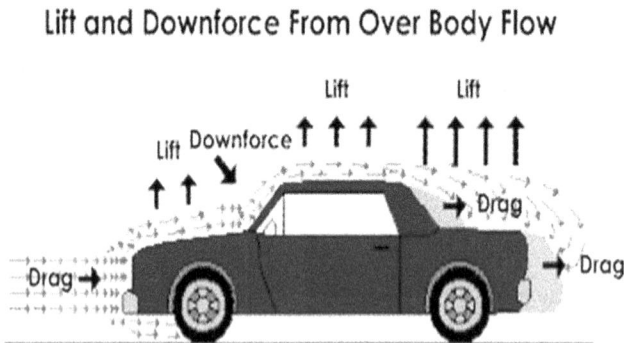

FIGURE 8.1 Drag force effects on a moving car.

FIGURE 8.2 Effect of spoiler on a moving car.

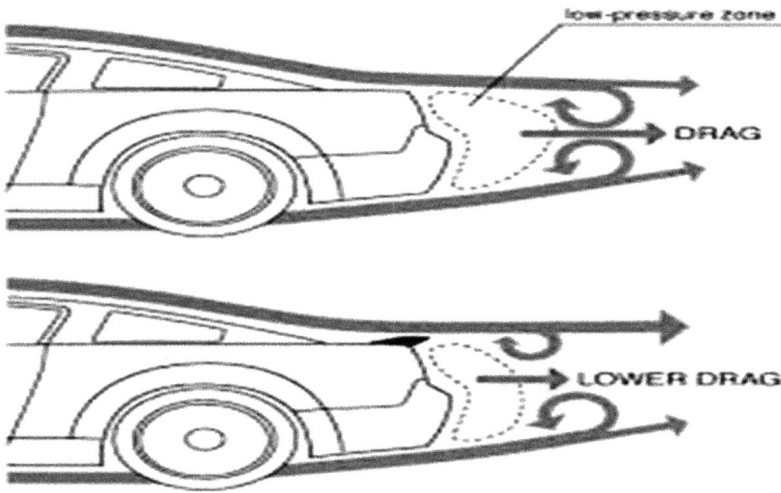

FIGURE 8.3 Effect of drag force by diffuser.

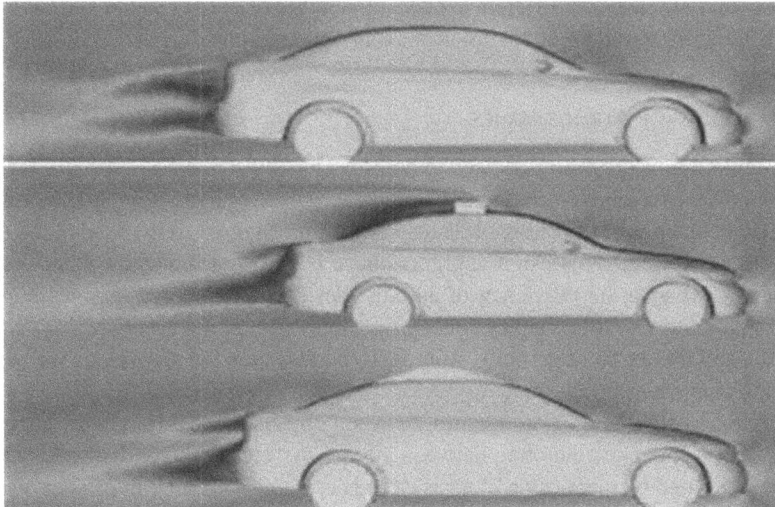

FIGURE 8.4 Effects of siren on a moving car.

Aerodynamic drag is one of the major obstacles while accelerating a body in the air. When a racing car or vehicle burns fuel to accelerate, drag force pulls the vehicle from back to reduce the speed and hence the fuel efficiency is adversely affected. Above 50%–60% of the total fuel energy is lost only to overcome the adverse aerodynamic force. Reduction of drag can be considered as a major area of study and learning and has become one of the prime concerns in vehicle aerodynamic nowadays.

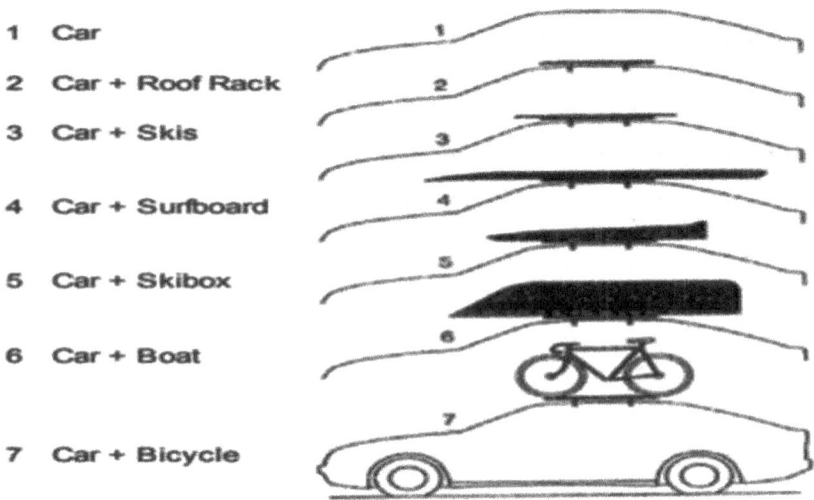

1	Car
2	Car + Roof Rack
3	Car + Skis
4	Car + Surfboard
5	Car + Skibox
6	Car + Boat
7	Car + Bicycle

FIGURE 8.5   Effects of roof rack on a moving car.

This study focuses on different aspects of analysis of drag of sedan and SUV cars and different drag reduction techniques using CFD methodology.

8.1.1 SCOPE OF AERODYNAMICS

The regulation of greenhouse gases to control global warming and rapidly increasing fuel prices have given tremendous pressure on the design engineers to enhance the current designs of the automobile using minimal changes in the shapes. To fulfill the above requirements, design engineers have been using the concepts of aerodynamics to enhance the efficiency of automobiles. Although aerodynamics depends on so many factors, this chapter concentrates on external devices, which affect the flow around the automobile body, to reduce the resistance of the vehicle in normal working conditions.

As we know about the aerodynamics (When objects move through air, forces are generated by the relative motion between air and surfaces of the body, study of these forces generated by the motion of air is called aerodynamics). In our project, various aerodynamic effects due to different changes in design and methodology of spoilers are considered. CFD being one of the efficient methodologies to determine the results is given high importance in our study.

Aerodynamic drag is one of the major obstacles while accelerating a body in the air. When a racing car or vehicle burns fuel to accelerate, drag force pulls the vehicle from back to reduce the speed and hence the fuel efficiency is adversely affected. Above 50%–60% of the total fuel energy is lost only to overcome the adverse aerodynamic force reduction of drag can be considered as a major area of study and learning and has become one of the prime concern in vehicle aerodynamic nowadays.

8.2 PROBLEM STATEMENT

The contemporary industry is facing certain aerodynamic issues that tend to decrease the safety, efficiency, and ride quality of a vehicle, which are given as follows:

1. Drag (fuel consumption, top speed, acceleration)
2. High-speed stability (lift)
3. Cross-wind stability (side forces and yawing moment)
4. Aero-acoustics (limiting the strength of sources).

8.3 INTRODUCTION TO SPOILER

A spoiler is an automotive aerodynamic device, intended design function of which is to "spoil" unfavorable air movement across the body of a vehicle in motion, usually described as drag. Spoilers on the front of a vehicle are often called air dams, because in addition to directing air flow, they also reduce the amount of air flowing underneath the vehicle, which generally reduces aerodynamic lift and drag. They are often fitted to race and high-performance sports cars, although they have become common on passenger vehicles as well. Some spoilers are added to cars primarily for styling purposes and have either little aerodynamic benefit or even make the aerodynamics worse. The goal of many spoilers used in passenger vehicles is to reduce drag and increase fuel efficiency.

Passenger vehicles can be equipped with front and rear spoilers. Front spoilers, found beneath the bumper, are mainly used to decrease the amount of air going underneath the vehicle to reduce the drag coefficient and lift [1,13].

8.3.1 INFERENCES DRAWN FROM PREVIOUS RESEARCH WORKS

CFD methodology has been developing now and then. For drag force prediction on passenger car with rear-mounted spoiler, this methodology can be adopted. The effect of spoilers can be studied experimentally through wind tunnel testing and analytically using CFD models. Drag prediction methodology can be carried out on different car models. In this study, we have considered a sedan and a SUV car designed model for performing CFD analysis although wind tunnel testing is significant in major automobile giant company and has its own uncompromised and essential benefits, but at the same time, performing a research or a study in a cost- and time-efficient manner becomes a constraint, consequently leading to the preference of CFD analysis. It is found that current trend in the industry is inclined towards analytical approach as time and cost associated with wind tunnel testing is very high [2].

The different types of roof racks are attached on passenger vehicles to carry luggage which affects aerodynamic drag. The objective of this work is to perform aerodynamic analysis of ground vehicle with a roof rack to investigate the change in drag coefficient by using CAD/CAM software. The aerodynamic analysis of a baseline passenger car model is performed with and without generic benchmarked roof rack at 100 km/h. Further analysis is carried out with different new designs of roof racks.

The result of the experiment is compared with the simulation result of scaled model which gives a variation of 12% (compared with CFD and experimental results).

The turbulence modeling effects on the CFD predictions of flow over a NASCAR Gen 6 race car were studied using three commonly used eddy viscosity turbulence models, realizable, AKN k-ε, and SST k-ω. The simulations were completed using a finite volume code with an unstructured predominantly hexahedral or trimmed mesh of 115 million cells. The prediction disagreements between different turbulence models are highlighted with delta drag and lift force plots along the vehicle model, and generated delta scalar scenes of pressure and velocity fields. The observed differences in the predicted flow-fields were explained in terms of differences in the vertical flow fields educed using the Q-criterion. All force coefficients will be presented in a non-dimensional form. Figure 8.6 represents the NASCAR Gen-6 cup race car model.

Parametric assessments of sedan car aerodynamic devices are presented into two groups of studies. These are rear end spoiler and underbody diffuser. The simplified design of these add-ons focuses on the main parameters such as length, position, or incidence leading to easier manufacturing for experiments and implementation in computational studies. The work is done on the CAD. Some different angle and shape are occupied for changing or lower the C_d. Figure 8.7 shows different views of the modeled vehicles.

The present evolution of the aerodynamics properties of 25% scale DrivAer model (in CFD and wind tunnel) including aerodynamic drag and lift is studied, and detailed investigations of the flow field around the vehicle is also done. In addition to the available geometries of the DrivAer model, individual changes can be introduced by modifying the geometry of the baseline model, and hence, total drag and lift values are reduced substantially. An individual subsection highlights the influence of cooling air flow (drag and lift), different rear ends, and underbody details, as well as geometry modifications. Figure 8.8 represents the simulations on vehicles.

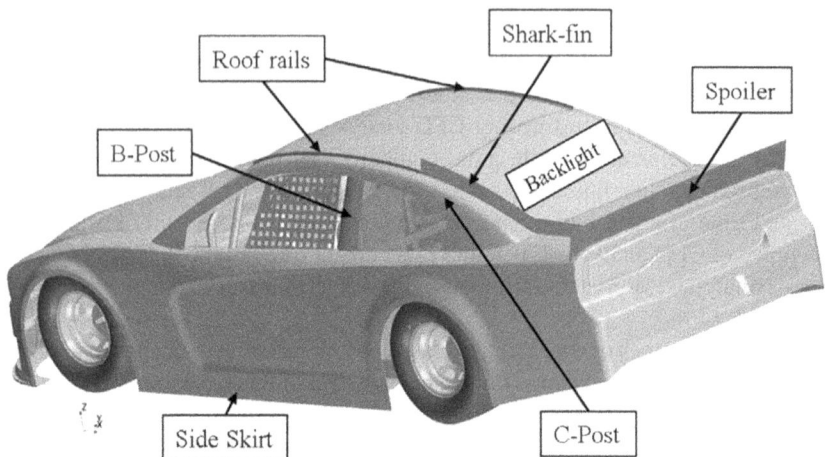

FIGURE 8.6   NASCAR Gen-6 cup race car model [3].

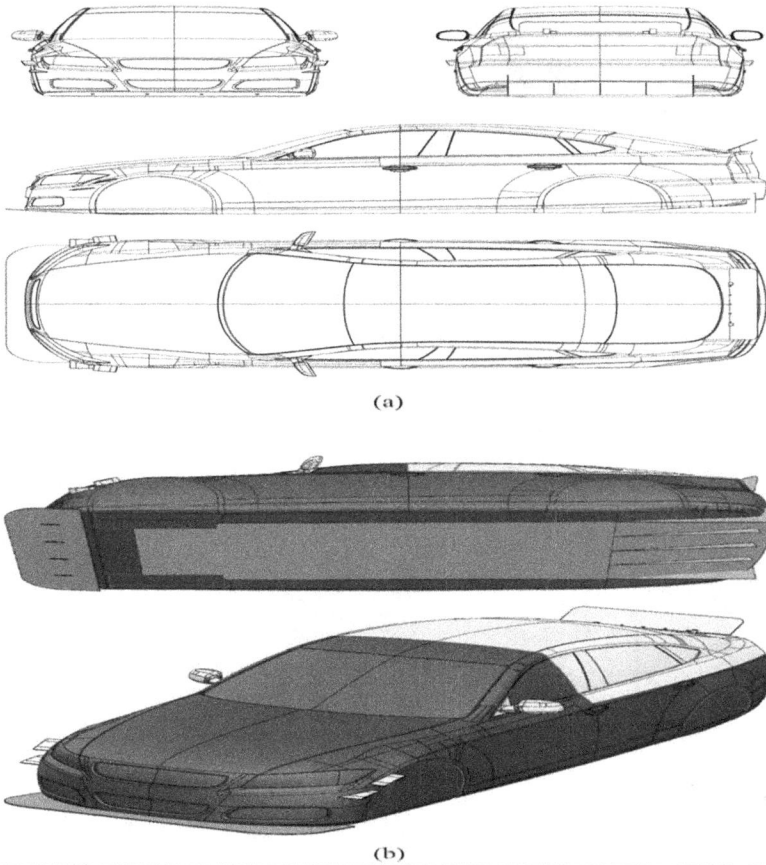

FIGURE 8.7 DrivAer hp-Fastback configuration: (a) projected, standard views; (b) perspective views showing the lower (splitter and diffuser) and upper (strakes and spoiler) vehicle model details [4].

Based on the design, developments and Numerical calculation of the effects of the external device, which will be spoiler that mounted at the rear side of the vehicle to make the present vehicles more aerodynamically attractive. The influence of rear spoiler on the generated lift, drag, and pressure distributions is investigated. This work is carried out using CFD. From previous investigations, it is found that 12° spoiler inclination angle model is the most optimum although it produces 1.56% extra C_D than 4° inclination angle. Minimum C_L maintained in the model is the basic concern for better stability of high-speed vehicle.

From previous studies about the active aerodynamic wings, investigations are carried out using numerical simulation in order to improve vehicle handling performance under emergency scenarios, such as tight cornering maneuver at high speeds. Figures 8.9 and 8.10 show the drawing of vehicles. This whole process was performed on CFD. Air foils are selected and analyzed to determine the basic geometric features of aerodynamic wings. Based on the airfoil analysis, the 3D aerodynamic

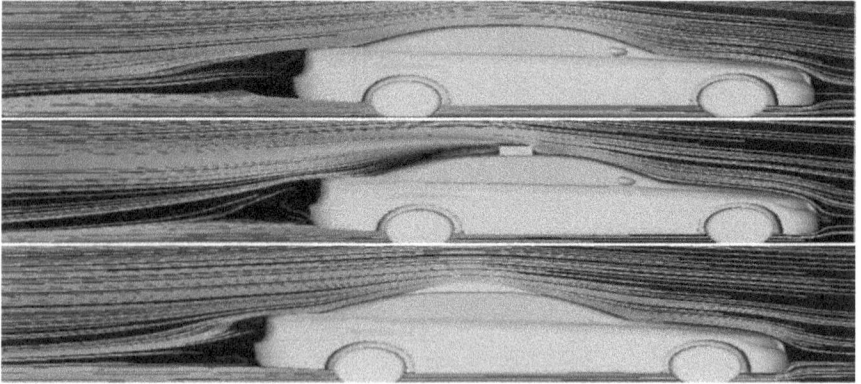

FIGURE 8.8 Velocity streamlines for original (a) with light bar (b) and optimum design (c) [5].

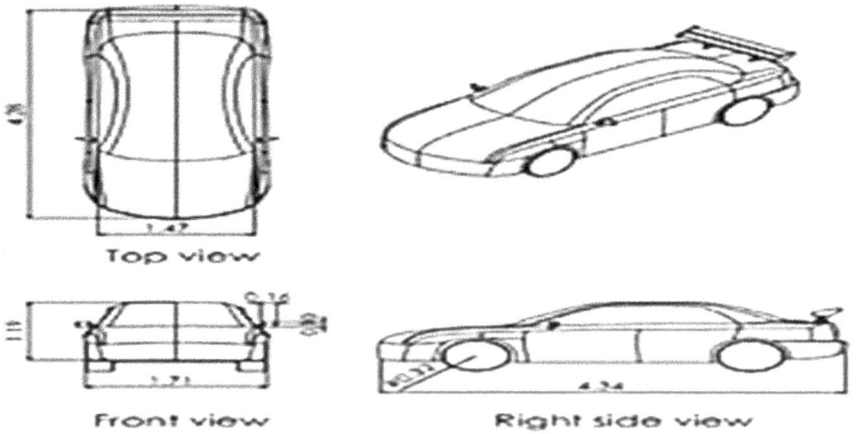

Top view

Front view

Right side view

FIGURE 8.9 3D vehicle model with relevant dimensions (meter) [6].

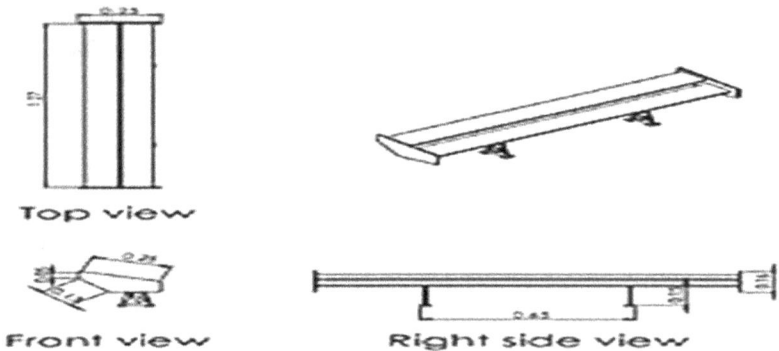

Top view

Front view

Right side view

FIGURE 8.10 Spoiler model with relevant dimensions (meter) [6].

wing model is developed using a commercial software package or airfoil shape used in spoiler to reduce C_D. Then, the virtual aerodynamic wings are assembled with the 3D vehicle model. The active aerodynamic control system with four independent wings is studied for improving the lateral stability of high-speed vehicles. The appropriate wing shapes are selected.

Instead of using very simple shape of a model, a more relevant simple car-like shape resembling a hatchback vehicle is utilized. This shape of a simple car body has a diffuser angle of $10°$. To give side tapering to it, the hatchback model is modified digitally. The boat-tailing is restricted to sides of a vehicle, and tapering is applied along the whole height of a vehicle. This study shows that changing the diffuser angle from $0°$ to $15°$ degree will affect the coefficient of drag. Figures 8.11 and 8.12 show the CAD model [7].

A 2D vehicle geometry of a race car is created and solved using the CFD solver FLUENT version 6.3. The aerodynamic effects are analyzed under various vehicle speeds with and without a rear spoiler. The main results are compared to a wind tunnel experiment conducted with 1/18 replica of a NASCAR. By the CFD analysis,

FIGURE 8.11   Vehicle model without aerodynamic wings [8].

FIGURE 8.12   Vehicle model with rear aerodynamic wing [8].

the drag coefficient without the spoiler is calculated to be 0.31. When the spoiler is added to the geometry, the drag coefficient increases to 0.36. The computational results with the spoiler are compared with the experimental data, and a good agreement is obtained within a 5.8% error band.

From previous studies, it came to picture that there is a considerable impact of vehicle add-ons on energy consumption and greenhouse gas emissions. Wind tunnel study was undertaken using a replica of reduced scale model passenger car. The aerodynamic drag for different add-on configurations was measured for a range of vehicle operating speeds and yaw angles. Various commonly used vehicle add-ons (police siren, advertising sign, taxi sign, and roof rack) and roof load (e.g., ladder and barrel) produce additional aerodynamic drag. The results show that the vehicle add-ons have notable impact on aerodynamic drag as they can generate 5%–40% more aerodynamic drag depending on cross-wind effects.

It has been studied that about 50%–60% of total fuel energy is lost only to overcome this adverse aerodynamic force which indicates aerodynamics of a vehicle has a great role to play to control different forces directed to it. Changes such as rear under-body modification and exhaust gas redirection towards the rear separation zones are made. A numerical process (finite volume method) of solving the Favre-averaged Navier–Stokes equations backed by k–epsilon turbulence model is carried out, the drag coefficient of the car under analysis is found, and it becomes evident from this study that there is a scope of research to reduce the drag further. Different rear under-body modifications and exhaust gas redirection towards the separated region at the rear of the car can be considered as a methodology to obtain the positive results, and consequently, the negative pressure area and its intensity at the rear of the car can be minimized and the separation pressure drag is subsequently reduced [9–10].

From previous studies about computational aero-acoustic analysis of a passenger car with a rear spoiler, a method is adopted where the topology of the test vehicle and grid system is constructed by a commercial package, ICEM/CFD FLUENT, which is a solver used in this study. After numerical iterations are completed, the aerodynamic data and detailed complicated flow structure are visualized using commercial packages: Field View and Tech plot. An effective numerical model is obtained based on the CFD approach to obtain the flow structure around a passenger car with wing-type rear spoiler. It is found that the installation of a spoiler with an appropriate angle of attack can reduce the aerodynamic C_L. Furthermore, the installation of an endplate can reduce the noise behind the car. Figures 8.13 and 8.14 show the simplified 2D model of a sedan car [11–12].

8.4   PROPOSED DESIGN OF A MOVING VEHICLE FOR CARRYING OUT FURTHER STUDY

8.5   CONCLUSION

1. Spoiler plays a crucial role in determining drag reduction in any vehicle, previous studies has deal with the changes in position of a spoiler on any vehicle. In our study drag reduction is estimated due to effect of height and

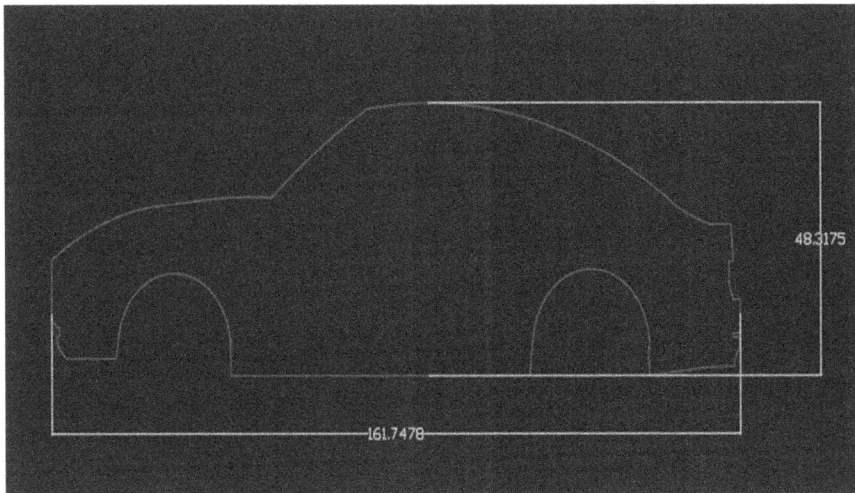

FIGURE 8.13 Simplified 2D model of sedan car (BMW 316 i).

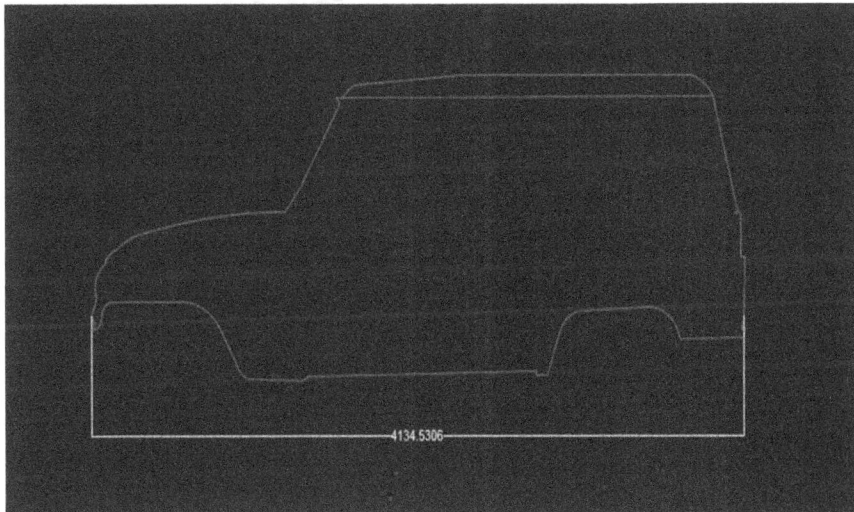

FIGURE 8.14 Simplified 2D model of a SUV car (Mahindra-BOLERO).

angle of a rear end spoiler specifically with flip-flap technique taking in consideration.

2. Diffuser also plays a vital role in study of drag reduction of any vehicle. Earlier studies do tells us about the position of diffuser but on the contrary, in our study we are trying to shorten the research gap by studying the drag reduction by changing both front and rear diffuser angle and also considering optimum spoiler height into action.

3. Cars having roof are common and the drag reduction effects due to it have already been a part of studies. In our research we are trying to reduce the research gap and making it more specific by also considering both with and without or convertible roof cars having the spoiler. The results of drag reduction due to it are including in our study.

4. Siren over cars specifically cop/police and ambulance is also a part of aerodynamics study. Research has shown that it plays a subsequent role in study of drag reduction. In our study here we are trying to be more specific and detailed to present the study of drag reduction due to effects of sirens on top of cars.

5. Certain add-ons like, ski roof/roof rack are place on the top of cars for various purposes. The effects of these on drag reduction of cars have been already a part of study. In our study, here we are making an attempt by going more specific and detailed. An SUV card will be taken and as an add-on ski roof and roof rack both are considered. The change in geometry of these add-ons is also one factor that is to be taken care of during study process and an attempt is made to present the complete study of drag reduction due to it.

REFERENCES

1. Pachpund, S., Madhavan, J., Pandit, G., "Development of CFD methodology for drag force prediction on passenger car with rear mounted spoiler," SAE International 2012–28–0029, 2019.
2. Hol, P. A., Agrewale, M. R., "Aerodynamic analysis of passenger car with luggage carrier (roof rack)," SAE Technical Paper 2019–26–0067, 2019.
3. Fu, C., Uddin, M., Robinson, A. C., "Turbulence modeling effects on the CFD predictions of flow over a NASCAR Gen 6 racecar," *Journal of Wind Engineering & Industrial Aerodynamics* 176 (2018) 98–111.
4. Soares, R. F., Knowles, A., Olives, S. G., Garry, K. et al., "On the aerodynamics of an enclosed-wheel racing car: an assessment and proposal of add-on devices for a fourth, high-performance configuration of the DrivAer model," SAE Technical Paper 2018-01-0725, 2018.
5. John, M., Buga, S.-D., Monti, I., Kuthada, T. et al., "Experimental and numerical study of the DrivAer model aerodynamics," SAE Technical Paper 2018-01-0741, 2018.
6. Taherkhani, A. R., Gilkeson, C., PhD Thesis, "Aerodynamic CFD based optimization of police car using bezier curves," SAE Technical Paper-10.4271/2017-01-9450, 2017.
7. Das, R. C., Riyad, M., "CFD analysis of passenger vehicle at various angle of rear end spoiler," *Procedia Engineering*, 194 (2017) 160–165.
8. Cai, J., Kapoor, S., Sikder, T., and He, Y., "Effects of active aerodynamic wings on handling performance of high-speed vehicles," SAE Technical Paper 2017-01-1592, 2017.
9. Palaskar, P., "Effect of side taper on aerodynamics drag of a simple body shape with diffuser and without diffuser," SAE Technical Paper 2016-01-1621, 2016.
10. Sadettin, H., El-Emam, R. S., "Effects of rear spoilers on ground vehicle aerodynamic drag," *International Journal of Numerical Methods for Heat & Fluid Flow* 24 (2014) 627–642.
11. Hassan, S. R., Islam, T., Ali, M., Islam, M. Q., "Numerical study on aerodynamic drag reduction of racing cars," *Procedia Engineering* 90 (2014) 308–313.

12. Chowdhury, H., Alam, F., Khan, I., Djamovski, V., Watkins, S., "Impact of vehicle add-ons on energy consumption and greenhouse gas emissions," *Procedia Engineering* 90 (2014) 308–313.
13. Tsai, C.-H., Fu, L.-M., Tai, C.-H., Huang, Y.-L., Leong, J.-C., "Computational aero-acoustic analysis of a passenger car with a rear spoiler," *Applied Mathematical Modelling* 33 (2009) 3661–3673.

9 Role of Finite Element Analysis in Customized Design of Kid's Orthotic Product

Harish Kumar Banga, Parveen Kalra, and R. M. Belokar
Punjab Engineering College

Rajesh Kumar
Panjab University

CONTENTS

9.1   INTRODUCTION

An ankle foot orthosis (AFO) is a mechanical device worn by drop foot patients with paretic ankle dorsiflexor muscles, to support and improve the working of the foot and ankle joint [1]. Although the aim of AFO is preventing the forefoot from drop in sway by obstructing the ankle movement, it also improves the ankle ability to support body weight, provides succession, and secures push-off ability for the period of stance phase of walking [2]. A lower leg foot orthosis (AFO) is usually utilized for foot drop brought about by lasting neuropathy. Customary assembling of AFO comprises manual mortar throwing, trim of thermoplastic materials, and cutting them as a type of AFO, which needs sensitive capacity and much effort [3]. Furthermore, the entire procedure of this assembling must be rehashed if the AFO is annihilated or a patient's condition is changed [4]. Three-dimensional (3D) printing procedure, otherwise called added substance fabricating, has been generally utilized in remedial fields, and their utilization is increasing violently [5–7]. Three-dimensional printers can deliver effectively modifiable articles with no fixed mouldings that make the things unmistakable. This 3D printing strategy makes it possible for doctors and specialists to make just patient-custom-made item for themselves. As of late, numerous preliminaries to produce Yes, an AFO with 3D printing procedure are finished. An orthosis made with 3D printing procedures has favourable circumstances in less sensitive expertise and exertion to make and simple multiplication over expectedly produced orthosis made by trimming the thermoplastic material [8–9]. Also, on the grounds that the planned 3D demonstrating file is spared once, assembling of an AFO can be handily rehashed. Moreover, if a programmed PC code program for of orthosis is created utilizing the pre-customized orthotic existing structure, the creation of the orthosis would be effortlessly accomplished and can be modified by patients for themselves [10–11].

    We build up a plan of orthosis, fabricated the orthosis, and made the orthosis utilizing the 3D printing procedure. At the present time, depicted this gathering strategy of changing over the past solid arrangement of AFO into two new parts and getting the parts along with a segment.

9.1.1   Passive Ankle Foot Orthosis

A latest designed AFO helps the patient by forestalling unfortunate foot movement during the swing period of the walk. Basically, it acts like a torsional spring, constraining the foot diversion by giving an outer torque [12]. Passive AFOs are economically accessible gadgets ordinarily utilized by persistence in daily strolling. Commercialized AFOs must be lightweight, durable, compact, and relatively inexpensive [13]. For commercial success and acceptance, AFOs could be either articulated or non-articulated devices, and are also categorized based on their constituent material which might be metal, leather, thermoplastic, composite, or a combination of materials in hybrid AFOs [14,15]. Thermoplastic AFOs, as shown in Figure 9.1, are the most widely recognized and are made as L-formed polypropylene plastic props, with the upstanding bit behind the calf and the lower divide under the foot. It is joined to the calf with a tie and is made to fit inside the accommodative shoes. In a thermoplastic AFO also called back leaf spring AFO, movement control attributes are determined by the material properties and the geometry [16,17] to modify joint consistence or damping.

Although rigid AFOs control motion during swing and inhibit the drop foot, they disturb normal motion behaviour in stance phase when foot is to plantar flex. Articulated thermoplastic and carbon fibre AFOs with commercial hinge joints such as tamarack flexure joint (Figure 9.2) are introduced to address this issue. It is worth noting that some joints such as tamarack also provide self-co-aligning characteristics of the medial and lateral joint axes [18].

Hybrid AFOs are developed in order to provide motion control in swing without unrestricted range of motion during stance. These devices consist of lightweight thermoplastic or carbon braces with articulated joints and passive elements aimed at storing and releasing for motion control [19]. A passive hybrid AFO called

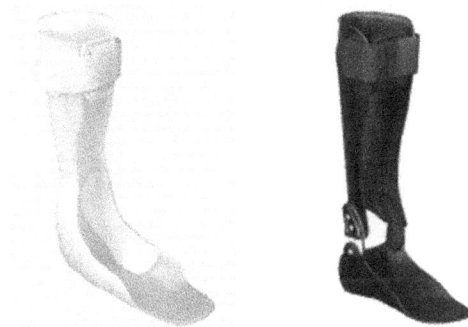

FIGURE 9.1 Passive thermoplastic AFOs: (a) plastic solid ankle AFO and (b) plastic articulating AFO.

FIGURE 9.2 Tamarack flexure joint for thermoplastic and carbon laminate bracing.

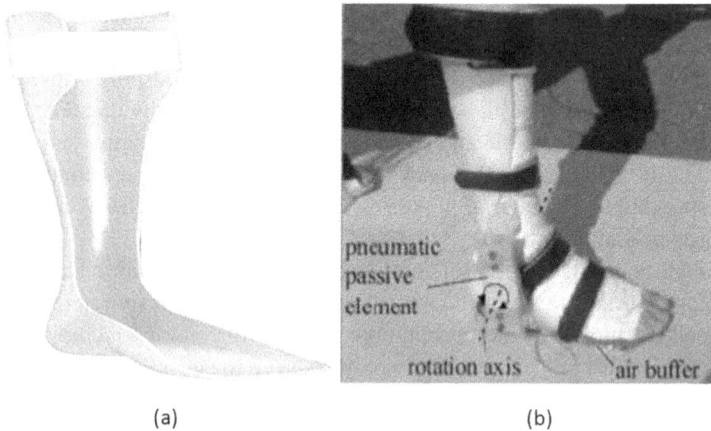

(a)   (b)

FIGURE 9.3   Passive hybrid AFOs: (a) the DACS's hybrid AFO and (b) the Osaka University AFO.

dorsiflexion assist controlled by spring (DACS) [20] is developed at University of Health and Welfare in Japan for drop foot patients (Figure 9.3a).

Also another hybrid device developed in Japan by using a passive pneumatic element to control the ankle joint (Figure 9.3b). These crossover AFOs empower the patient to utilize an assortment of shoes and furthermore give biomechanical alternatives to customizability of the lower leg joint [21].

9.1.2 ACTIVE ANKLE FOOT ORTHOSIS

As discussed earlier, detached AFOs produce over the top protection from plantar flexion in position stage. Along these lines, specialists have been keen on dynamic AFOs, expected to modify impedance of the orthotic joint for different bits of the stride cycle [22]. All the previously mentioned dynamic AFOs (Figure 9.5) should be associated with the outside force supplies and PCs for activity. In this manner, the relevance of these orthoses is right now restricted to research facility contemplates. A versatile fuelled lower leg foot orthosis (PPAFO) was created at the College of Illinois [23], utilizing a pneumatic actuator, a CO_2 power source, and a locally available controller. In spite of the fact that this AFO presents an unfastened controllable gadget, execution of the gadget as a supported restoration apparatus for day-by-day wearing relies upon future examinations and improving the continuance of the AFO (Figure 9.4).

9.1.3 TYPES OF AFOS

There is an assortment of orthosis for various body parts and signs. Sometimes, an orthosis is also called sidebars or sidebar/band orthosis. In the orthotic field, FIOR and GENTZ is represented considerable authority in the lower limit. As a rule, there is a qualification between orthosis for patients with loss of motion and alleviation

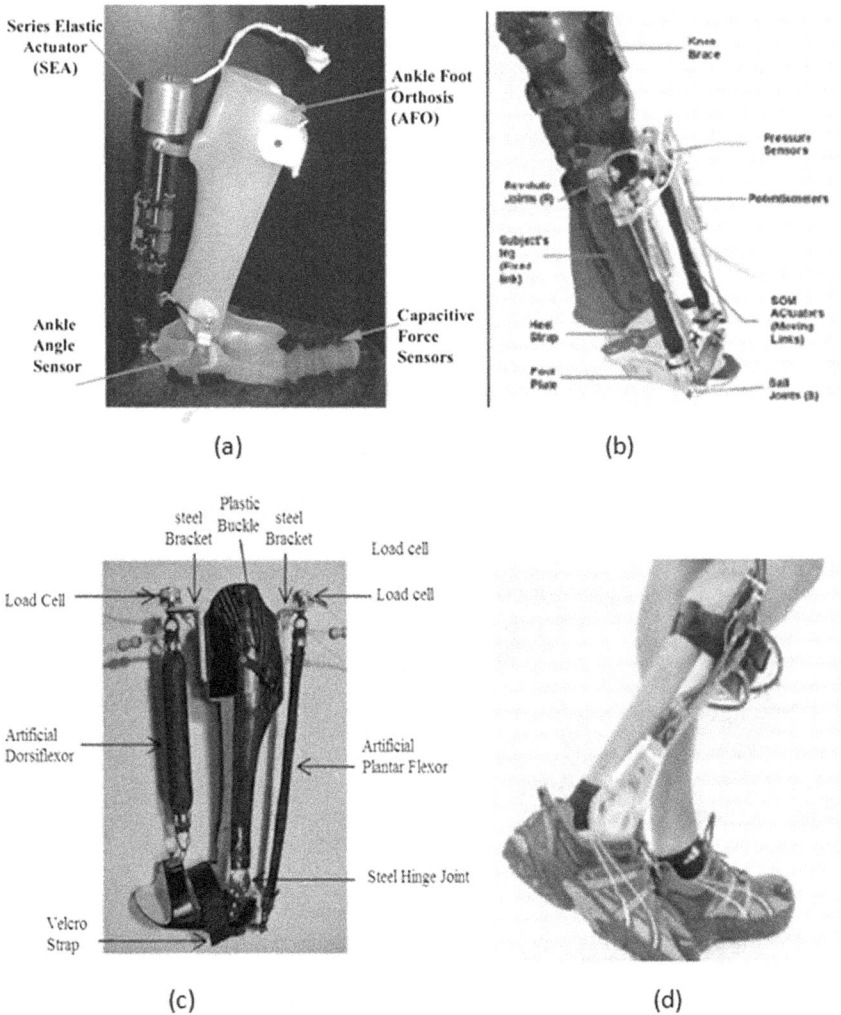

FIGURE 9.4 Active AFOs: (a) MIT's SEA active AFO; (b) the University of Michigan AFO powered by pneumatic muscles; (c) the Arizona State University robotic tendon AFO; (d) the University of Champaign Illinois PPAFO.

orthosis. In addition, we discussed various sorts of orthoses assuming a significant job right now [24].

9.1.3.1   Ankle Foot Orthosis (AFO)

An AFO is also called lower leg orthosis or lower leg foot orthosis. The joint in the orthosis expect the capacity of the lower leg joint or those capacities the foot can't perform without anyone else any longer (lifting) [25].

An AFO goes up to the knee probably; however, it doesn't cover it. Contingent upon the necessities, an AFO is either recreated or uniquely crafted. The potential

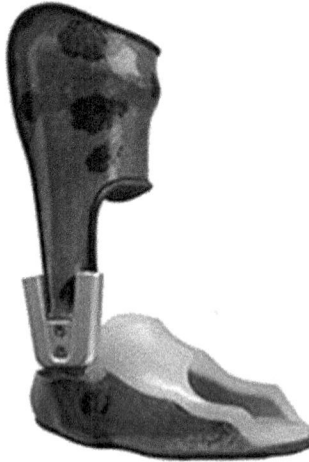

FIGURE 9.5   Ankle foot orthosis.

outcomes of the orthosis joint are of definitive significance. It can have a foot lifting impact, impacting the dorsiflexion (augmentation of the foot towards the tibia) and the plantar flexion (flexion of the foot towards the ground) [26].

9.1.3.2   Knee Ankle Foot Orthosis (KAFO)

A knee ankle foot orthosis (KAFO) extends over the knee. The focus is on the joint used at knee height (Figure 9.6).

They are divided into free moving, locked, and automatic system knee joints. The joint is always selected according to the patient's indications.

9.1.3.3   Knee Orthosis (KO)

A knee orthosis (KO) only covers the knee. Contrary to a KAFO or an AFO, there is no joint at ankle height (Figure 9.7).

Knee orthosis are used (post-operatively) in order to protect and relieve the knee joint. Thus, knee pain can be relieved considerably.

Frequent indications are, e.g., cruciate ligament injuries, capsular ligaments injuries, and meniscus injuries.

9.1.3.4   Foot Orthosis (FO)

Indication is

- Arch support in patients who are prorated (flat feet) (Figure 9.8).

9.1.3.5   University of California-Berkeley Lab (UCBL)

Indications are

- Arch and heel support
- Higher level of support than foot orthoses (Figure 9.9a).

FIGURE 9.6 Knee ankle foot orthosis.

FIGURE 9.7 Knee orthosis.

9.1.3.6 Supra-Malleolar Orthosis (SMO)
- Can help control proration (flat feet) AND suppuration (arches too high)
- Relatively rigid control, but ankle and knee motions still allowed (Figure 9.9b).

9.1.3.7 Posterior Leaf AFO
- Controls foot drop (toe pointing down)
- Allows the ankle to come forward when the foot is on the ground during walking
- Helps push foot off the floor before it swings through (Figure 9.10).

FIGURE 9.8   Foot orthosis.

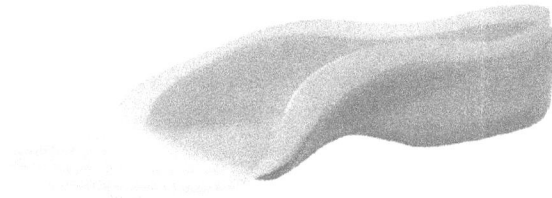

FIGURE 9.9A   University of California-Berkeley Lab (UCBL).

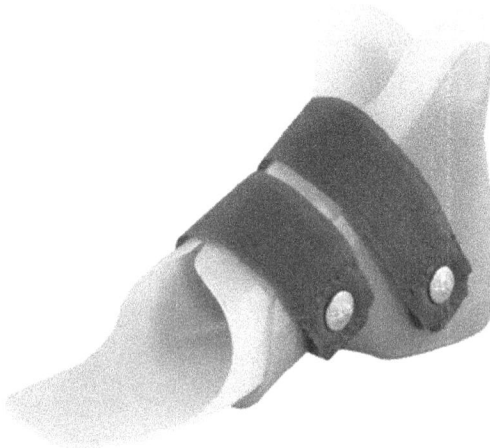

FIGURE 9.9B   Supra-malleolar orthosis (SMO).

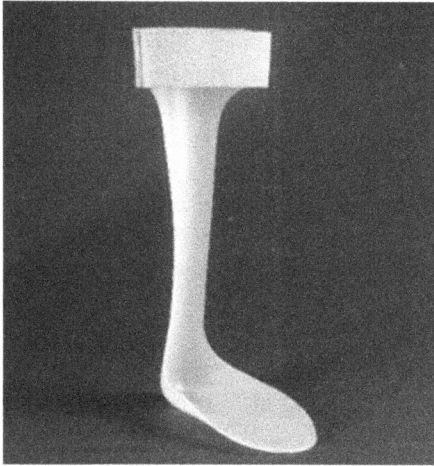

FIGURE 9.10 Posterior leaf spring AFO.

9.1.4 BENEFITS OF AFO

- Improve or prevent a physical deformity
- Stabilize a joint or joints
- Reduce pain
- Improve mobility or performance
- Reduce the risk of tripping over
- Reduce the risk of injury.

One approach to help keep this sort of strolling is to fit a beneath knee lower leg/foot orthosis (AFO) which can help control any strange development of the foot and lower leg amid strolling, play or rest.

- A well-made and tight AFO will help balance out the foot and lower leg to realize lower leg dependability and improve parity, stance, and certainty.
- An AFO can help lift the foot, forestalling stumbling over, diminishing mishaps, and making strolling less demanding and less tiring.
- Controlling the foot and lower leg will likewise impact hip and knee position in a positive way and thus lead to potential upgrades in the youngster's walk, parity, and stance.
- The foot plate of the AFO can be leveled or moulded relying upon the tyke's necessities. Insoles can be joined into the AFO to help keep up a decent foot pose.
- Modifications can likewise be made to the outside of the AFO or even the kid's footwear subsequent to fitting, so as to "adjust" the AFO. Little, post-supply modifications can likewise expand the viability of the AFO; this is most usually completed to alter heel stature, which changes the point of the lower leg in connection to the ground when strolling (tibial shaft edge).

- An AFO ought to almost certainly fit into the youngster's very own footwear (not generally issue free but rather typically conceivable), which implies they will be eager to wear it as well. A few guardians have anyway answered to Hemi Help that more extensive shoes or potentially bigger shoes have been important to oblige an AFO.
- An AFO ought to never be awkward for a kid to wear, and kids ought to probably wear it cheerfully for the majority of their dynamic day. Once in a while, AFOs can rub on weight centers and make "sensitive areas", so it's vital to screen your tyke's foot, particularly the impact point and lower leg.

9.2 METHODOLOGY

A three-dimensional mesh geometry directly from patients' limb was taken using the 3D laser scanner. Consequently, 3D printable orthotic design is created from simple input model by means of SOLIDWORKS software. A customized design is produced by breaking the solid piece in two different parts. To make these parts work, they were joined with a joining mechanism, and eventually, the two mechanisms were analysed on ANSYS, and the best mechanism is to be 3D printed (Figure 9.11).

9.2.1 PROBLEM DEFINITION

First, we are going to design the mechanism for our AFO which are of two types: sliding type and locking type. Sliding and locking mechanisms of AFO have their own glitches. Hence, throughout the course of this project, we have to study both mechanisms thoroughly and decide which will be best after analysing various designs under design software like CAD and analysing the application of forces on it in ANSYS.

9.2.2 DESIGN OF THE NEW MECHANISM

The solid model previously given to us had problems: less flexibility, more rigidity, less stress/strain endurance capacity, and difficulty in motion. To overcome all these problems, we break this model into two parts, namely, leg and foot, and use some

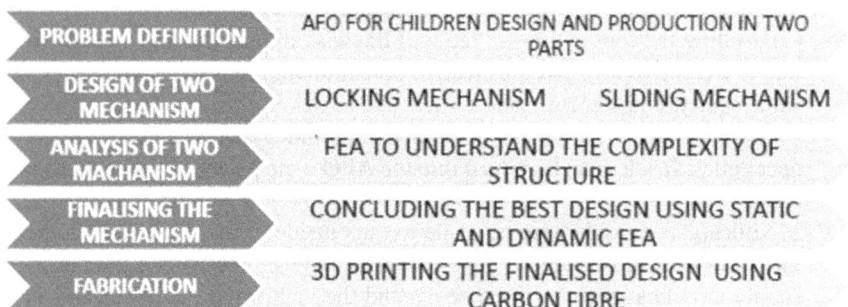

FIGURE 9.11 Flow diagram for kids AFO design.

linking mechanisms for joining and simultaneously overcome the above problems. Hence, we suggested two mechanisms: adjustable and hinged. After studying the two mechanisms in detail, we proposed a design for the two mechanisms, which are discussed next (Figures 9.12 and 9.13).

Moreover, in the stress analysis of the designs, we found out that the benefits of our proposed design came out to be more flexible and higher stress endurance, and less deformation under weight of child. Material used gives us better results in all tested fields against polypropylene and hence has high impact strength. The material now used has better surface finish and better design durability.

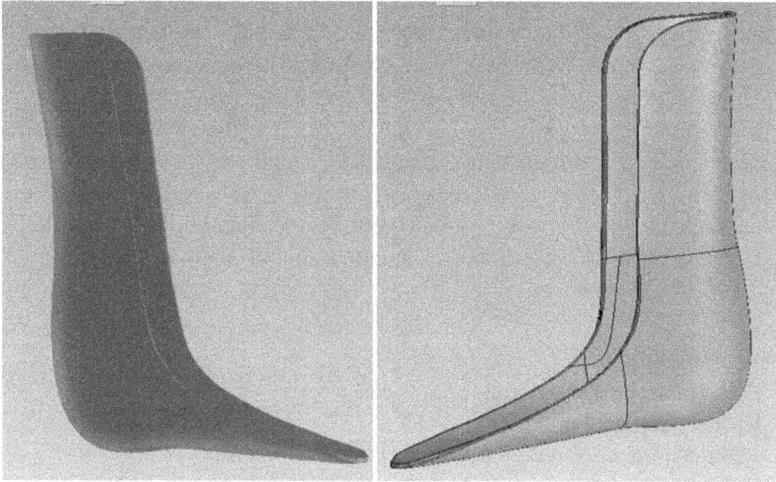

FIGURE 9.12   Existing design of kids AFO.

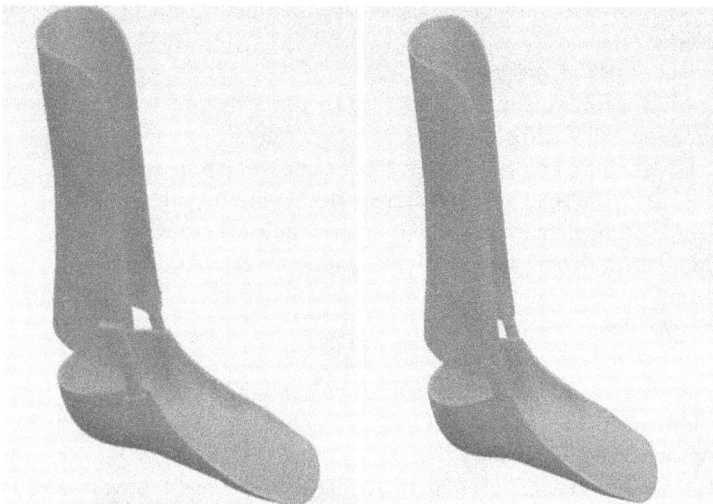

FIGURE 9.13   Customized design of kids AFO.

9.2.3   Design Analysis

We managed to compile data of various weights of kids ranging from 15 kg (min.) to 20 kg (max). Hence, we have to analysed the data for different weights.

9.2.3.1   Basic Principle

Consider a frame or designing element through which the appropriation of a discipline variable, e.g., relocation or strain, is needed. Precedents may be a section beneath burden, temperatures difficulty to a warmth enter, and so forth. The body, for instance, a one- or a few-dimensional sturdy, is displayed as being theoretically subdivided right into a get together of little elements referred to as components – "constrained additives". The phrase "restricted" is applied to painting the confined, or constrained, variety of levels of possibility used to demonstrate the conduct of every thing. The additives are notion to be associated with every other, however, simply at interconnected joints, called hubs. Note that the additives are notionally little locales, not separate substances like blocks, and there are no breaks or surfaces between them.

The manner towards speaking to a component as a gathering of confined components, known as discretization, is a standout among the most important strides in the FEM of examination.

9.2.3.2   Advantages

- Performance improvement for new and existing items
- Reduce plan and assembling costs
- Reduce number of physical test preliminaries required
- Bring items to showcase quicker
- Possibility to assess and streamline elective structures and materials
- Quickly dissect variations of essential arrangements
- Better structure of item because of clear comprehension of the plan aim of parts
- Reduce material wastage
- Topology and weight enhancement of the structure without influencing execution and well-being
- Better structure choices conceivable because of complete data
- Reach to conclusive item structures with required quality quicker
- Meet legitimate and legally binding prerequisites effectively
- Gain credit from perceived confirmation experts, locally just as all around.
- Improve item security
- Optimize the structure against dynamic vibrations
- Determine the item conduct in genuine conditions by applying fitting limit conditions on the model
- Improve consumer loyalty
- FEA can be executed even on moderate workstations and work area PCs
- FEA is presently an incorporated apparatus in the vast majority of the advanced CAD bundles.

9.2.4 FINITE ELEMENT ANALYSIS OF AFO

Finite element analysis (FEA) is crucial part as it describes different stress, strain, and forces acting on the AFO . It helps to analyse the complex structure and system, determine component behaviour and accurately predict how the product will react under stress.

ANSYS software has been used for FEA (Figures 9.14 and 9.15).

Before an ANSYS evaluation, the geometry of AFO is damaged up into small pieces called elements. The corner of every detail is a node. A node is a coordinate location in space in which the degrees of freedom (DOFs) are described. For

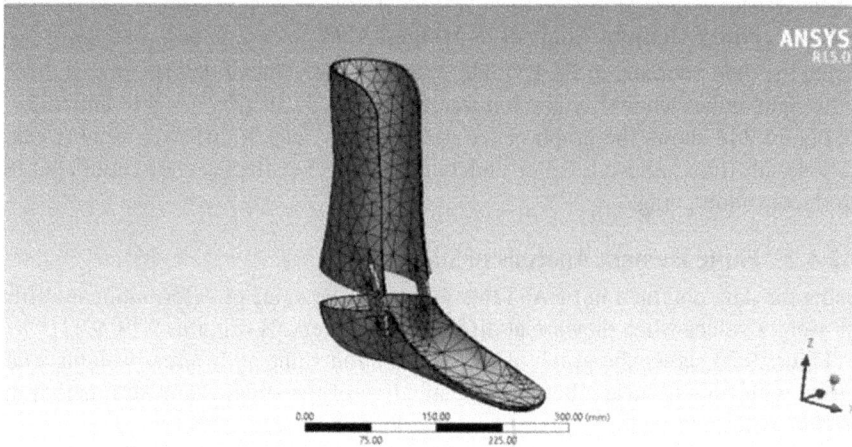

FIGURE 9.14 Meshing of adjustable AFO design-I.

FIGURE 9.15 The meshing and nodes in sliding AFO.

structural analyses, the DOFs represent the possible motion of the factor because of the loading of the shape. The pressure within the fabric is decided from the relative motion of the nodes, and the stresses are calculated primarily based at the strains and the fabric properties. The calculation is accomplished at the nodes. Additionally, nodes can be included on the midpoint alongside the edges of the elements. These elements and nodes make up the mesh. Solution accuracy depends on an excellent mesh, and ANSYS automates an awful lot of the mesh introduction system to assist create an awesome mesh for the simulation.

Table 9.1 gives the quantity of factors the AFO has been divided into and the subsequent number of nodes gift, and then the FEA is performed to calculate the pressure and pressure at the product.

9.2.4.1 Finite Element Analysis of Hinged AFO

Using the data obtained in FEA, Table 9.1 shows the value of deformation at different weight values when they are at different time intervals (Figures 9.16 and 9.17).

Figure 9.18 shows the graph of the deformation value at different weights when varied with time to have a better understanding of the stresses and strains that the product is undergoing.

9.2.4.2 Finite Element Analysis of Sliding AFO

Using the data obtained in FEA, Table 9.2 shows the value of deformation at different weight values when they are at different time intervals (Figures 9.19–9.22).

Figure 9.23 shows the graph of the deformation value at different weights when varied with time to have a better understanding of the stresses and strains that the product will go through.

- The deformation at all the feasible weights, obtained after FEA, is given in appendix.

TABLE 9.1
Deformation of Design with Different Weights

	Carbon Fibre Testing on Different Weight					
Time	15 kg	16 kg	17 kg	18 kg	19 kg	20 kg
0.1	0	0	0	0	0	0
0.2	0.132	0.141	0.15	0.159	0.168	0.178
0.3	0.265	0.283	0.3	0.318	0.336	0.357
0.4	0.368	0.424	0.451	0.477	0.504	0.536
0.5	0.53	0.566	0.601	0.637	0.672	0.715
0.6	0.663	0.707	0.752	0.796	0.84	0.893
0.7	0.793	0.864	0.902	0.955	1	1.072
0.8	0.929	0.991	1.05	1.115	1.176	1.251
0.9	1.061	1.132	1.203	1.274	1.344	1.43
1	1.194	1.274	1.353	1.433	1.513	1.608

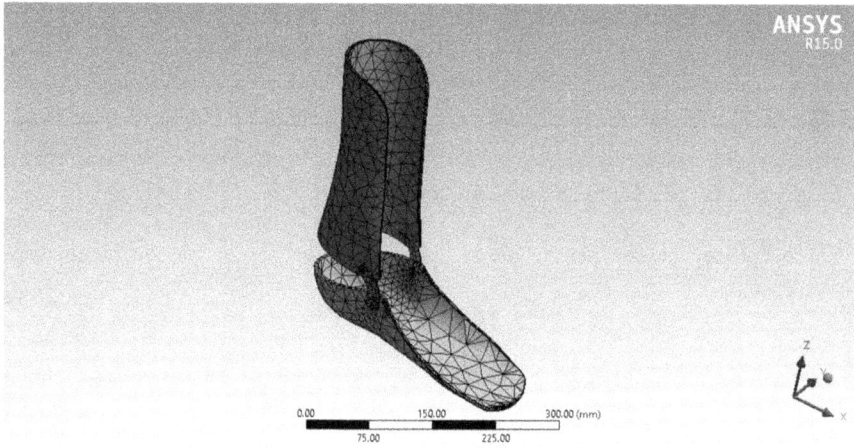

FIGURE 9.16   Meshing of adjustable AFO design-II.

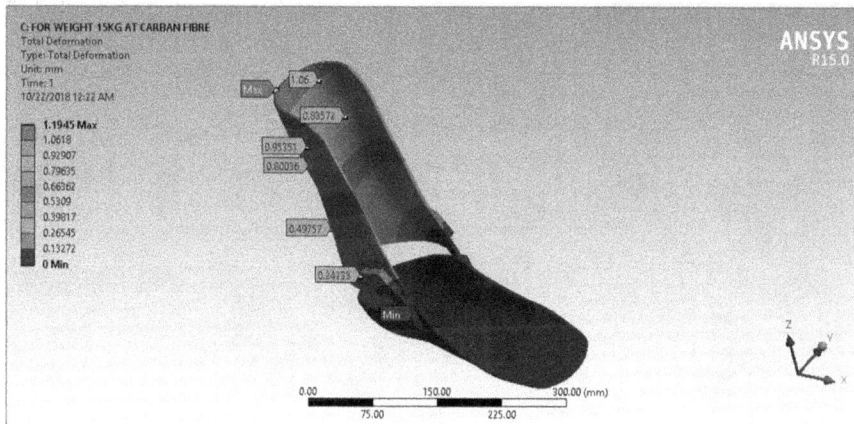

FIGURE 9.17   The deformations and stress new design.

Deformations for Sliding AFO.

The colour in the images highlights the intensity of the total deformation where dark colour shows the minimum deformation, and as the colour increases to light, the deformation also increases (Figure 9.24, Table 9.3).

9.2.5   Finalizing the Mechanism and Fabrication

Our next and the most important step is finalizing the design for fabrication. The design finalized should be a complete homonym of appropriate stress measured, lesser deformation, and more flexibility. Next step is making the physical object from the 3D model typically by laying down many thin layers of carbon fibre from the digital file.

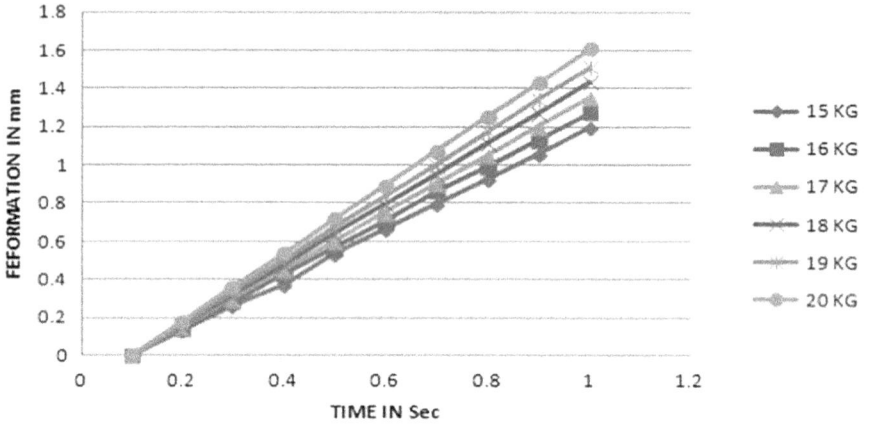

FIGURE 9.18 Graph between deformation vs. time of AFO design I.

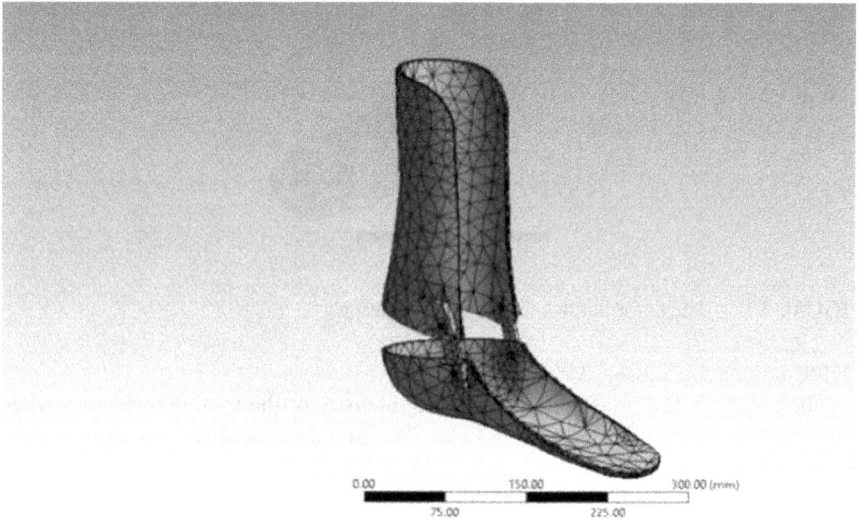

FIGURE 9.19 Meshing with nodes and element details of the sliding AFO.

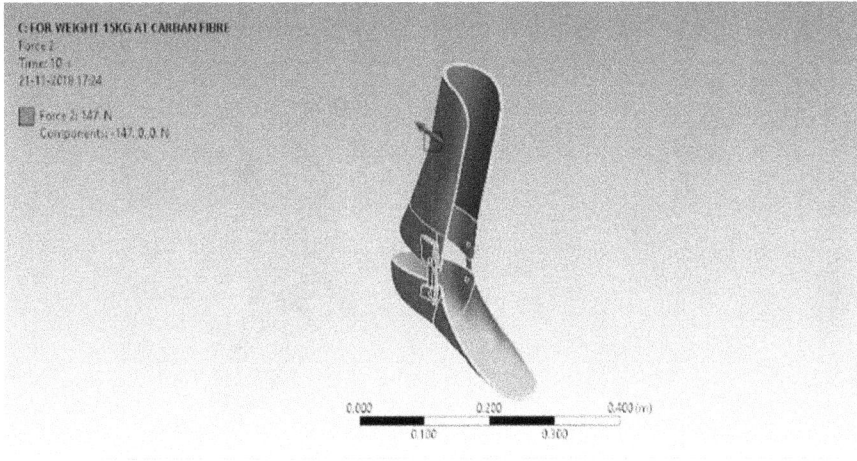

FIGURE 9.20 Force applied in AFO design.

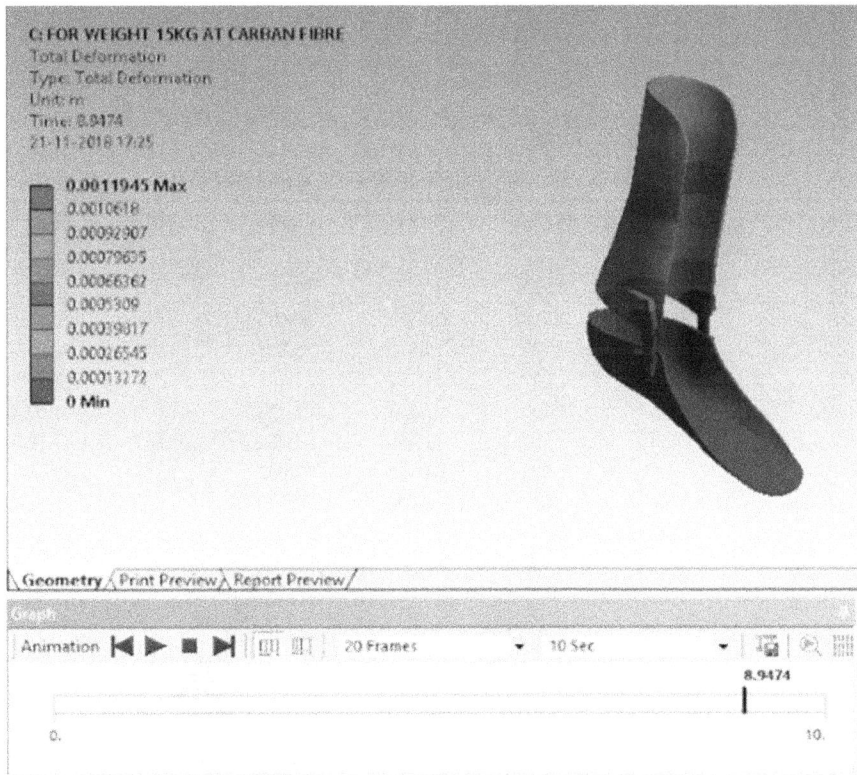

FIGURE 9.21 The deformation of AFO design.

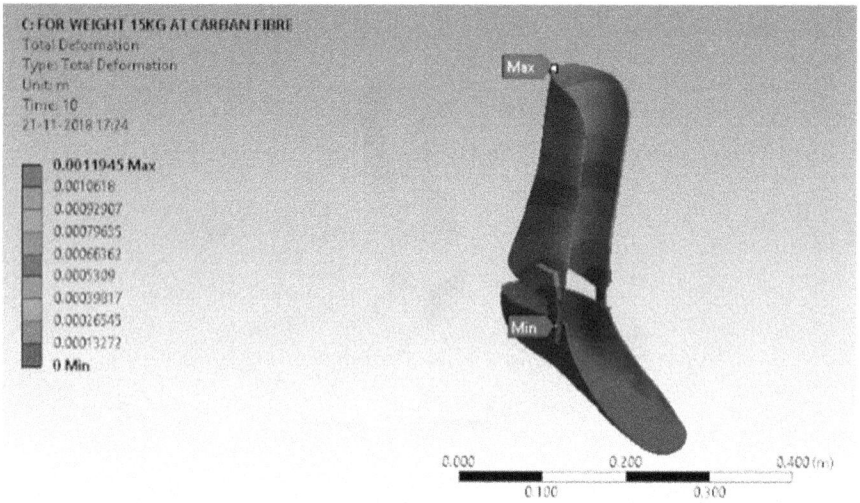

FIGURE 9.22   The stresses and deformations at maximum weight in detail at a weight of 15 kg.

TABLE 9.2
AFO Design-II Time vs. Load

	Carban Fibre Testing on Different Weight Model Design 2					
TIME	15 kg	16 kg	17 kg	18 kg	19 kg	20 kg
0.1	0	0	0	0	0	0
0.2	0.243	0.259	0.275	0.291	0.307	0.324
0.3	0.486	0.518	0.551	0.583	0.615	0.648
0.4	0.729	0.777	0.862	0.875	0.923	0.972
0.5	0.972	1.037	1.102	1.166	1.231	1.296
0.6	1.215	1.296	1.377	1.458	1.539	1.62
0.7	1.458	1.555	1.653	1.75	1.847	1.944
0.8	1.701	1.815	1.928	2.042	2.155	2.268
0.9	1.944	2.077	2.204	2.333	2.463	2.593
1	2.1878	2.337	2.479	2.625	2.771	2.917

Markforged software was used to fix areas which would contain nylon and carbon fibre. The reason behind taking two materials is that mixture of these materials gives a greater strength of model.

9.3   RESULTS

The arranged settled which ended up being the adjustable segment gives out the lesser distortion on utilization of intensity and bears near weight because of territory

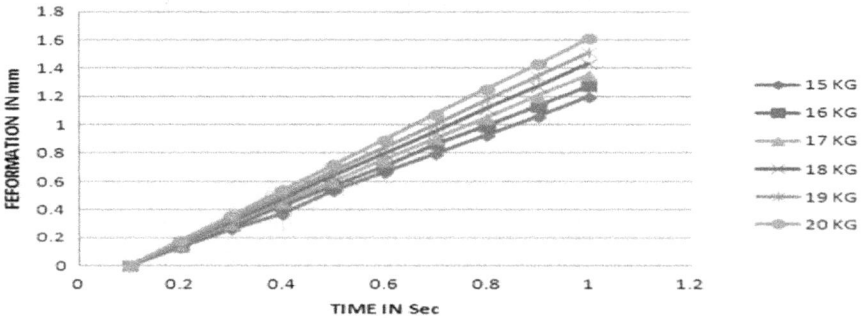

FIGURE 9.23 Graph between different weight and load in AFO design.

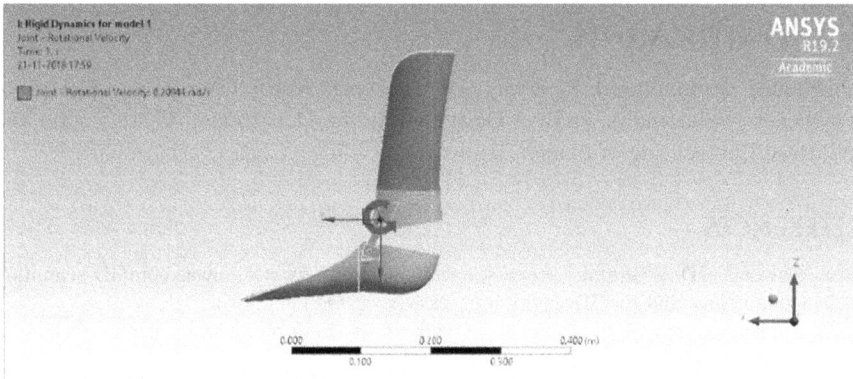

FIGURE 9.24 The model and sliding dynamics of the AFO model.

TABLE 9.3
The Measurement Units Used in the FEA

Unit System Metric	(m. kg, N, s. v, A) Degrees ran/s Celsius
Angle	Degrees
Rotational velocity	rad/s
Temperature	Celsius

under contemplations. This flexible system would be utilized for 3D printing using a Markforged 3D printer having onyx and carbon fibre as inks. These materials will invigorate out the vital adaptability and required for manufacture. The results of this examination indicated that the course of action of excessive sweating and warming of lower leg foot orthosis to patients having age bunch three to five years during changing the climate conditions for long utilization. The genuine part and another structure for patients in field-tested strategy programming and do the restricted segment assessment with the objective that we find which arrangement is closer to the genuine bit of AFO.

9.4　CONCLUSIONS

The aim of this research will be to fabricate the AFO. This will be observed by way of trying out on sufferers for gait evaluation ultimately the usage of it for mass manufacturing. Future works at the improvement of the AFO testbed will recognize at the layout of the clamp factors so that it will provide quantitative frame weight for the duration of the gait. Actual human gait cycle facts and ankle stiffness could be applied to the control system so that you can affirm the practical evaluation of the AFO. Further study of the energy return in gait will be analysed in this AFO to take a look at the bed. Control gadget will be evolved within the real-time device so that it will enhance the performance.

ACKNOWLEDGMENTS

This study is supported by Prof. (Dr.) Parveen Kalra Coordinator Centre of Excellence (Industrial & Product Design) from the Department of Production and Industrial Engineering of Punjab Engineering College, Chandigarh, India.

REFERENCES

1. Aniwaa, 3D scanning process, available: http://www.aniwaa.com/3D-scanning-technologies-and-the-3Dscanning-process (accessed Feb 2016).
2. Aniwaa, Artec Eva, available: http://www.aniwaa.com/product/3D-scanners/artec-eva (accessed Feb 2016).
3. Annabi, N., Tamayol, A., Uquillas, J. A., Akbari, M., Bertassoni, L. E., Cha, C., and Khademhosseini, A., "25th anniversary article: rational design and applications of hydrogels in regenerative medicine", *Advanced Materials*, 26(1), 85–124. 2014.
4. Apeagyei, P. R. "Application of 3D body scanning technology to human measurement for clothing fit", *Change*, 4(7). 2010.
5. Arbace, L., Sonnino, E., Callieri, M., Dellepiane, M., Fabbri, M., Idelson, A. I. and Scopigno, R. "Innovative uses of 3D digital technologies to assist the restoration of a fragmented terracotta statue", *Journal of Cultural Heritage*, 14(4), 332–345. 2013.
6. Banga, H. K., Kalra, P., Belokar, R. M. and Kumar, R. "Rapid prototyping applications in medical sciences", *International Journal of Emerging Technologies in Computational and Applied Sciences (IJETCAS)*, 5(8), 416–420. 2014.
7. Banga, H. K., Belokar, R. M., Madan, R. and Dhole, S. "Three dimensional Gait assessments during walking of healthy people and drop foot patients", *Defence Life Science Journal*, 2, 14–20. 2017.
8. Banga, H. K., Belokar, R. M., Kalra, P. and Madan, R. "Fabrication and stress analysis of ankle foot orthosis with additive manufacturing", *Rapid Prototyping Journal, Emerald Publishing, Rapid Prototyping Journal*, 24(1), 301–312. 2018.
9. Banga, H. K., Belokar, R. M. and Kumar, R. "A novel approach for ankle foot orthosis developed by three dimensional technologies", *3rd International Conference on Mechanical Engineering and Automation Science (ICMEAS 2017)*, University of Birmingham, UK, Vol. 8, No. 10, pp. 141–145. 2017.

10. Banga, H. K., Kalra, P., Belokar, R. M. and Kumar, R. "Fabrication and customized design of kids ankle foot orthosis by 3D printing", *2nd International conference on New frontiers in engineering, Science & technology (NFEST-2019) NIT*, Kurukshetra, India. 2019.

11. Banga, H. K., Kalra, P., Belokar, R. M. and Kumar, R. "Effect of 3D-printed ankle foot orthosis during walking of foot deformities patients". In: Kumar H. and Jain P. (eds) *Recent Advances in Mechanical Engineering. Lecture Notes in Mechanical Engineering.* Springer, Singapore. 2020.

12. Chae, M. P., Rozen, W. M., McMenamin, P. G., Findlay, M. W., Spychal, R. T. and Hunter-Smith, D. J. "Emerging applications of bedside 3D printing in plastic surgery", *Frontiers in Surgery*, 2, 25. 2015.

13. Chang, J. W., Park, S. A., Park, J. K., Choi, J. W., Kim, Y. S., Shin, Y. S. and Kim, C. H. "Tissue engineered tracheal reconstruction using three dimensionally printed artificial tracheal graft: preliminary report", *Artificial Organs*, 38(6), E95–E105. 2014.

14. Ciocca, L. and Scotti, R. "CAD-CAM generated ear cast by means of a laser scanner and rapid prototyping machine", *The Journal of Prosthetic Dentistry*, 92(6), 591–595. 2004.

15. Dubravčik, M. and Kender, Š. "Application of reverse engineering techniques in mechanics systemservices", *Procedia Engineering*, 48, 96–104. 1992.

16. Elmqvist, M., Jungert, E., Lantz, F., Persson, A. and Soderman, U. "Terrain modelling and analysis using laser scanner data", *International Archives of Photogrammetry Remote Sensing and Spatial Information Sciences*, 34(3/W4), 219–226. 2001.

17. Giannatsis, J. and Dedoussis, V. "Additive fabrication technologies applied to medicine and health care: a review", *The International Journal of Advanced Manufacturing Technology*, 40(1–2), 116–127. 2009.

18. Gibson, I. Kvan, T. and Wai Ming, L. "Rapid prototyping for architectural models", *Rapid Prototyping Journal*, 8(2), 91–95. 2002.

19. Herr, H. "Exoskeletons and orthosis: classification, design challenges and future directions", *Journal of Neuroengineering and Rehabilitation*, 6(1), 1. 2009.

20. www.alimed.com/afo.

21. Hoskins, S. "3D printing for artists, designers and makers", Bloomsbury. 2013.

22. Hull, C. W. U.S. Patent No. 4,575,330. Washington, DC: U.S. Patent and Trademark Office. 1986.

23. Istook, C. L. and Hwang, S. J. "3D body scanning systems with application to the apparel industry", *Journal of Fashion Marketing and Management: An International Journal*, 5(2), 120–132. 2001.

24. Jensen-Haxel, P. "3D printers, obsolete firearm supply controls, and the right to build self-defense weapons under Heller", *Golden Gate University Law Review*, 42, 447. 2011.

25. Jin, Y. A., Plott, J., Chen, R., Wensman, J. and Shih, A. "Additive manufacturing of custom orthosis and prostheses-a review", *Procedia CIRP*, 36, 199–204. 2015.

26. Joe Lopes, A. MacDonald, E. and Wicker, R. B. "Integrating stereolithography and direct print technologies for 3D structural electronics fabrication", *Rapid Prototyping Journal*, 18(2), 129–143. 2012.

10 An Experimental Study of Mechanical and Microstructural Properties of AA7075-T6 by Using UWFSW Process

Akash Sharma, V.K. Dwivedi, and Shahabuddin
GLA University

CONTENTS

10.1 INTRODUCTION: BACKGROUND OF WELDING

Welding can trace its historic development back to ancient times. The earliest examples come from the Bronze Age. Small gold circular boxes were made by pressure welding lap joints together. It is estimated that these boxes were made more than 2000 years ago. During the Iron Age, the Egyptians and people in the eastern Mediterranean area learned to weld pieces of iron together. Many tools were found which were made approximately 1000 BC. During the Middle Ages, the art of blacksmithing was developed, and many items of iron were produced which were welded by hammering.

Welding is the process by which two pieces of metal can be joined together. The process of welding doesn't merely bond the two pieces together as in brazing and soldering but, through the use of extreme heat and sometimes the addition of other metals or gases, causes the metallic structure of the two pieces to join together and become one. There are mainly six types of welding:

1. Forged welding
2. Thermit welding
3. Gas welding
4. Electric arc welding
5. Resistance welding
6. Friction stir welding.

Forged welding is an ancient welding process used by blacksmith to join two pieces by repeated hammering of the two pieces in red-hot condition. Thermit welding is suitable for remote places and bigger components where gas and electricity are not available. Gas welding is suitable for joining thin plates. Electric arc welding is the most common popular method for welding. Resistance welding is most commonly used to join thin sheets, e.g., welding of automobile bodies. Friction stir welding (FSW) is a process to join two similar or dissimilar metals by the rotating tool in which process utilized heat generated by the rotating tool as well as plastic deformation. In FSW, work pieces being welded were rigidly clamped together on the special design fixture, and a cylindrical tool having shoulder and pin is rotated with suitable speed and plunges inside two pieces until the shoulder make a contact with the surface of the base metal (BM). Friction heat is generated by rotating tool between work piece and rotating tool as shown in Figure 10.1.

Generated heat due to the rotating tool to soften the BM without reaching the melting point and traversing tool mixes the material, thereby completing the joint. The rotation tool generates heat between BM and shoulder diameter of the tool, softening the material and pushing it simultaneously from the advance side (AS) to the retreating side (RS) to complete the welding joint. Thus, the BM is mechanically mixed until it undergoes plastic deforming without melting [1,2].

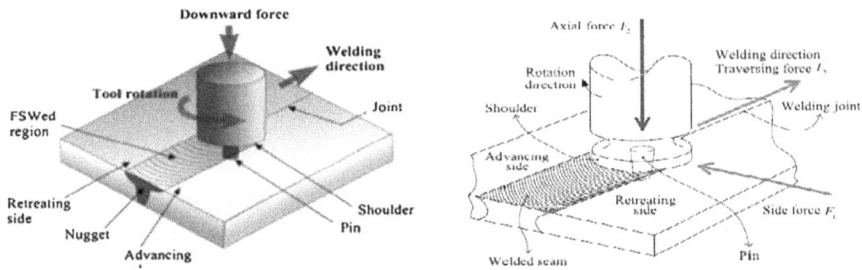

FIGURE 10.1   Schematic drawing of FSW process.

The tool shoulder prevents the flow material from the AS and RS to make better the welding bead and also obtain the better tensile strength quality of the joint. Plastic deformation of the BM due to the action of rotating tool with pin leads to dynamic recrystallization of the BM, which results in grain refinement to mechanical properties of the joint.

During FSW, BM moves around the pin without reaching the melting point. Generally in conventional fusion welding, difficulties arise during change of state, but not in this process [3].

Due to lower welding temperature, residual stresses are generated in FSW process, which results in improvement in fracture toughness and fatigue that make thin material welding possible. The FSW setup process is very costly compared to conventional welding process, and it requires less skilled operators, as shown in Figure 10.1.

A number of potential advantages of FSW over conventional fusion welding processes are listed as follows:

i. Good mechanical properties in the as-welded condition.
ii. Improved safety due to the absence of toxic fumes or the spatter of molten material.
iii. No consumables – a threaded pin made of conventional tool steel, e.g., hardened H13.
iv. No filler or gas shield is required for aluminum welding.

Moreover, FSW is considered as a green and environmentally friendly welding technology because of low energy consumption, no gas emission, and no need for consumable material such as electrodes, filler metals, and shielding gases (normally present in fusion welding processes) [4,5]. A survey carried out by the American Welding Society (AWS) in 2002 showed that $34.4 billion per year is spent on arc welding including the use of consumables, repair, and energy consumption in the USA. The adoption of FSW has increased rapidly, and 10% of joining processes have reportedly been replaced by FSW.

In the FSW process, the material to be welded is properly tightened with the help of fixture and mounted on the vertical milling machine as shown in Figures 10.7 and 10.8 to complete the welding joint in which BM is fixed in such way that the slide plate properly guides the BM and does not lift the BM during process. Fixture design

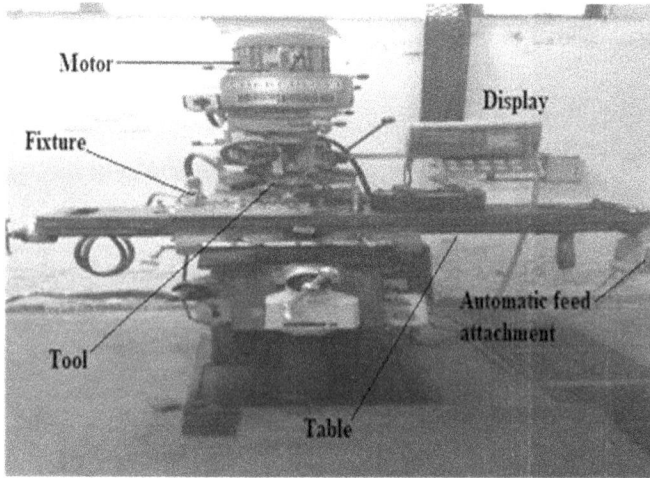

plays a very important role in making a conventional welding joint [6,7]. FSW process can be performed in different sequence to complete the weld joint.

10.2 FRICTION STIR WELDING

FSW is a latest method of bonding of two similar dissimilar metals with the rotational tool. This process was developed by the Welding Institute (TWI) of UK in 1991 [8,9]. FSW is also called a solid-state joining operation of two lightweight metals with lower melting point in such way that increased tensile strength of the weld joints with increasing friction time. All the friction-welded joints failed at the friction interface in tensile test. FSW quickly received the attention of many researchers around the world. In FSW process, a joint is made by a rotational tool which is inserted into work pieces to be joined and moved beside the weld joint [9]. FSW is an innovative and emerging welding process and material processing technique now extensively used in products of aluminum industries. The technology has gained increasing interest and improvement since its invention [8]. The basic principle of FSW is explained, and the continuing development of the technology is described from the perspective of discovery, invention, and innovation. Particular attentions will be given to the mechanical and structure integrity that can be expected from FSW technology.

In this section, a comprehensive literature review for experimental and analytical study of FSW on aluminum alloys under different welding parameters is presented. The review is related to the effect of tool geometry, design of holding devices, i.e., fixtures, effect on mechanical properties of welded joint, study of heat affected zone (HAZ), and changes in microstructure of the material after FSW.

10.3 TOOL GEOMETRY

N.D Ghetiya and K M Patel et al. (2015) observed that in FSW process with a cylindrical tool which contains pin and shoulder, tool parameters also play an important role in determining joint's characteristics. The FSW process makes a plastic flow and frictional heat so that it may be regarded as mechanical process. FSW welding is a continuous process to join the two metals by a non-consumable rotating tool on base materials. P. Ganesh and V. S. Senthil Kumar (2015) from experimental results indicate that tool rotational speed is the critical parameter during FSW that has a greater effect on super plastic formations. Hence, super plasticity can be significantly enhanced by FSW process due to variations in FSW tool rotational speed.

Joaquín M. Piccini and Hernán G. Svoboda (2017) analyzed the mechanical properties of the welded joints by micro-hardness profiles and by peel and cross-tension tests. The fracture loads increased when the tool penetrates inside the work piece. The tool geometry optimization increased the fracture loads. R. Rai et al. (2011) studied the rotating tool and focusing effect on mechanical properties and different tool shapes and rotational speed by using FSW as shown in Figure 10.2. Therefore,

FIGURE 10.2   (a) Cylindrical threaded, (b) three-flat threaded, (c) triangular, (d) Trivex, (e) threaded conical, (f) schematic of a triflute.

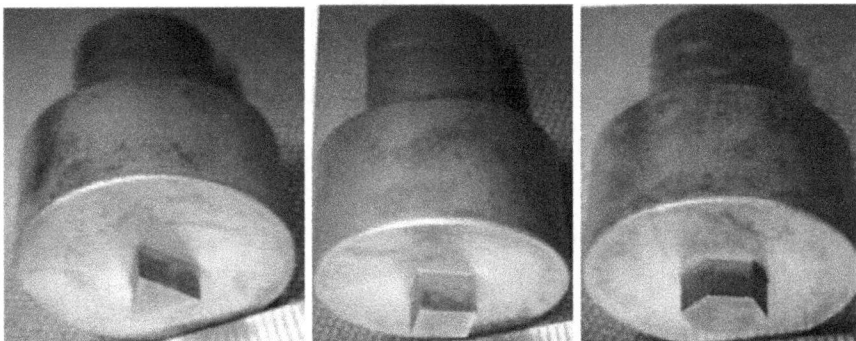

FIGURE 10.3   Polygonal tool pin profiles: constant dynamic volume covered by pin: (a) triangular, (b) square, and (c) hexagonal.

in this study, it has been tried to understand the tool shapes and parameters play an important role to determine the tensile strength of weld joints by FSW.

Kush P. Mehta and Vishvesh J. Badheka (2016) reported maximum irregular and large copper particles in welds made by triangular pin profiles. Polygonal pin shape has defects such as fragmental cracks and voids irrespective of its static and dynamic areas. The defects decreased with the increase in polygonal. Defect-free macrojoint was reported for cylindrical tool pin profile.

10.4 MATERIAL SELECTION

High-strength aluminum alloys are used in automobile, aircraft, and aerospace because of their excellent strength as shown in Table 10.1 & Table 10.2. So, FSW process can be considered as the most efficient way to join the aluminum alloy material [11,12]. Lower processing temperatures along with better efficiency make it a very demandable and robust technique for joining two facing work pieces. During FSW precipitation-hardened aluminum alloys, it is usually reported that the HAZ is developed due to the coarsening of the precipitate. AA7075-T6 alloys have high-strength materials for structural applications in aerospace field. AA7075-T6 aluminum alloy material with a 6 mm thick plate (AA7075) is selected for experimental work. Work piece is used in this size 150 mm×125 mm×6 mm for experiment work in FSW process as shown in Figure 10.4. This experiment has been performed in the Department of Mechanical Engineering (University Polytechnic), Jamia Millia Islamia (a central university), New Delhi; a vertical milling machine is used to perform FSW process. A non-consumable (FSW) tool with different shapes and sizes made of En-31 is used to perform the weld joints at different speeds by FSW. The

TABLE 10.1

Chemical Composition of AA7075 according to Spectrometer Analysis (wt.%)

Element	Fe	Si	Cu	Mn	Al	Mg	Zn	Cr	Ti
Required wt.%	0.4	0.3	1.2–2.0	0.3	87.1–90.2	2.0–2.8	5.0–66.2	0.190–0.29	0.3

TABLE 10.2

Mechanical Property of AA7075-T6 according to Spectrometer Analysis

Brinell hardness	150
Yield strength (MPA)	572
Ultimate tensile strength (MPa)	503
% Elongation	11
Modulus of elasticity (MPa)	70.7
Fracture toughness (J)	20
Poisson ratio	0.33
Machinability	70%

shoulder diameter of tool of 20 mm, front pin diameter of 5 mm and pin length of 5.75 mm are used for FSW process. In FSW, rotating tool generates heat, which results in local plastic deformation [1,2].

10.5 PREPARATION OF SPECIMENS FOR FSW

Aluminum alloy AA7075-T6 materials were purchased as plates from Suresh Metal, , Mumbai, India. Samples were prepared to a size of 62 mm×180 mm×6 mm by using shaper machine and hacksaw for FSW process as shown in Figure 10.4.

10.6 CHARACTERISTICS OF FRICTION STIR WELDING

FSW produced better weld quality with the following characteristics:

Low distortion: In FSW process, the plate distortion was performed in FSW machine. The test was carried out on an aluminum plate with sideway bends smaller than 0.25 mm and thinner metal, slightly bent in upward direction.

Low shrinkage: FSW produced the same amount of shrinkage in FSW process and in every time found lower than wide aluminum alloy panel application.

No porosity: There is no porosity in the weld joint because the base metal does not melt.

No lack of fusion: This is a forging and extruding joining process in which heat is more accurately controlled, so there is no lack of fusion in the joints.

No change in material: In FSW process, 450°C heat is generated during this process; no additional filler metal and heat is required to the joint. Due to the fine grain in nugget zone, the joint formed is stronger than in the BM.

10.7 UNDERWATER FRICTION STIR WELDING

Underwater friction stir welding (UWFSW) is a new technique for joining different aluminum alloy series such as 6xxx, 7xxx, and 8xxx, which are lightweight and have high strength which is widely acceptable in the industries like aerospace and

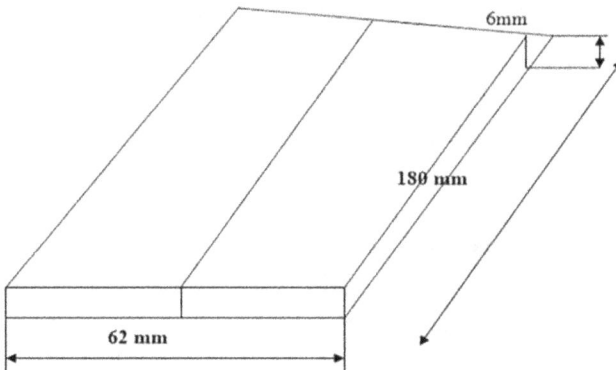

FIGURE 10.4 Dimensions of the weld joint specimen for FSW process of AA7075.

FIGURE 10.5   USFSW process.

shipbuilding. It is very difficult to weld these alloys by using fusion method so, FSW is introduced and it is widely taken into consideration for performing such welding process. Here, aluminum alloy AA7075-T6 is used for performing the experimental work. It is having good mechanical properties, higher strength, and lighter in weight and used for the fabrication of different body parts of airplane. UWFSW is a process in which the mechanical properties are improved because of the water introduced as cooling medium which helps in cooling and due to which the overheating of HAZ is minimized , which results in better mechanical properties. In UWFSW, the strength of the welded joint is increased in air medium compared to FSW.

In USFSW, the separate setup of fixture is used for performing such process(Figure 10.5) so that the water can be filled and then the work piece is immersed in the water.

In UWFSW process, through the water flow on the surface of the workpiece the heat input can be easily controlled either in low level, medium level or high level of different heat inputs which directly results in the increase of tensile strength as well in improved micro-structure and grain size. [1–3]. The mechanical properties and micro-structure are improved in USFSW. Toughness of the specimen is decreased due to the process which is taking place under water. Intense heat generated during the USFSW plays a vital role in the formation of the fine grain in NZ and also the microstructure at thermo-mechanically affected zone (TMAZ). In USFSW process, the mechanical properties and microstructure are found better as compared to simple FSW process.

10.8   EXPERIMENTAL PROCEDURE

First, the weld specimen is clamped in the in-house designed fixture as shown in Figure 10.6. In USFSW process, the material used for performing experimentation is Aluminum Alloy AA7075-T6 due to its high strength and lightweight and its wide acceptance throughout the industries for the fabrication of lightweight structures and different components.

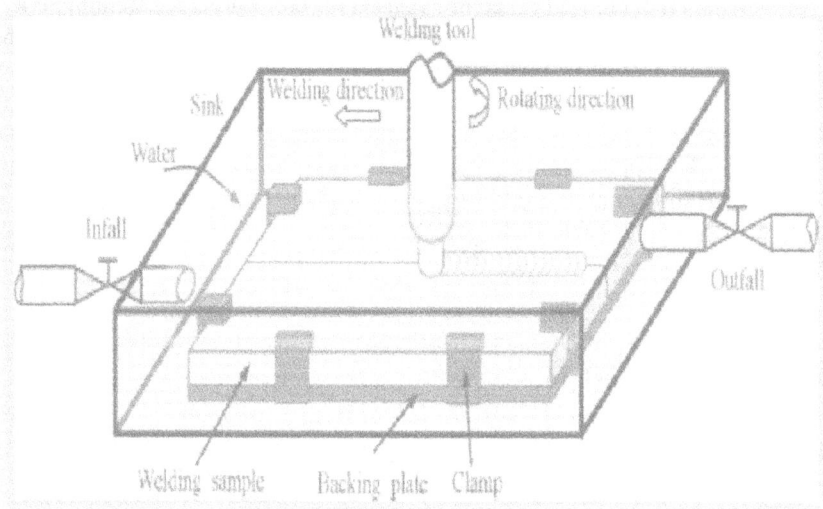

FIGURE 10.6 Fixture used for UWFSW process.

FIGURE 10.7 Vertical milling machine used for UWFSW.

The tool used UWFSW process is made of H13 steel. It is made by taking into consideration certain specified dimensions. UWFSW process tool specifications play very important role to maintain the mechanical properties and microstructure [10]. Here, the welding process is carried out by using H13 steel-made tool, chemical composition of which is shown in Table 10.3. In this process, a shoulder diameter of 20 mm and a pin diameter of 5 mm are used in the experimental research work as shown in Figures 10.2 and 10.2. A tool pin is very useful for filling the metal into the cavity.

A tool with a tool plunge of 0.20 mm is used for performing the welding process on a vertical milling machine. Work piece is firmly tightened on the special designed fixture for underwater FSW. First, specimen is placed on the fixture plate, and then, water is dropped, until the specimen is fully immerged into the water during the UWFSW process. While designing a tool, appropriate specification should be taken because the whole processes rely on the tool that plays a very important role in performing welding joint in UWFSW process. The different process parameters are shown in Table 10.3 & Table 10.4.

TABLE 10.3
Chemical Composition of H-13 Tool Steel

S. No	Element	Required wt.%
1	C	0.32
2	S	0.02
3	Mn	0.43
4	P	0.021
5	Si	1.05
6	Cr	4.98

TABLE 10.4
Process Parameters Used in UWFSW

Parameters	Levels (rpm)			
Tool rotational speed (N) (RPM)	900	1110	1200	1320
Welding speed (S) (mm/min)	55	60	65	70
Tool shoulder diameter (D) (mm)	20	20	20	20
Plunged depth (h) (mm)	0.10	0.12	0.15	0.20

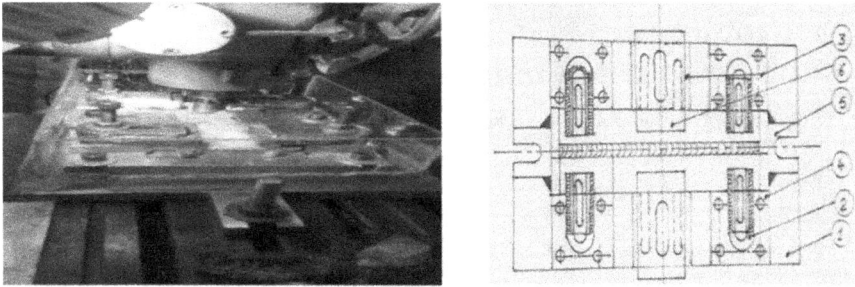

FIGURE 10.8 UWFSW with and without fixture setup.

10.9 DESIGN OF EXPERIMENT (DOE)

Design of experiment technique is used to perform UWFSW process to obtain better results in welded joint. Three input parameters, tool shoulder diameter, welding speed, and tool speed, were selected to perform the experiments in this research work to give better results when the number of factor is 3, with their levels shown in Table 10.4. Mostly, experiments were conducted to obtain the defect-free welding joints. In this research work, full factorial arrays with 3 columns and 16 rows were selected as shown in Table 10.5. The tensile tests were carried out in IIT Delhi, India, at a machine head speed of 2 mm/min.

Weld joint specimens were prepared using double-disc polished machine, and different reagents were used to examine the microstructure through a metallurgical microscope (Jamia Millia Islamia, New Delhi). Scanning electron microscope (SEM) (Jamia Millia Islamia, New Delhi, India) was used to produce the images of fracture features of different specimens. A Vickers micro-hardness testing machine (HM200, Mitutoyo and Kawasaki, Japan) was used for micro-hardness measurement of the welded specimens by applying a load of 2 N and a delay time of 20 s. Successive distance between two points was 0.5 mm.

In this research work, different parameters such as shoulder diameter, plunge depth, and tool speed were used for UWFSW. These parameters were varying by Taguchi L16 orthogonal array. Taguchi L16 array method was very suitable tool for engineering analysis with varying parameters at different levels to minimize the number of experiment trials.

10.10 TENSILE STRENGTH

In this research work, analysis of tensile strength is carried out due to the effect of different parameters in UWFSW. Tensile strength is maximum at 1750 rpm/min (435 MPa) and minimum at 900 rpm/min. Ultimate tensile strength (UTS) was observed experimentally. Taguchi method with support of ANOVA was used to analyze the mode of fracture study. It is observed that the second level of rotational tool speed (1110 rpm), the second level of welding speed (70 mm/min), and the third level of tool shoulder diameter of 20 mm (396 MPa) give higher tensile strength.

TABLE 10.5
L16 Orthogonal Array Design of Experiment (DOE)

S. No	Tool Rotational Speed	Plunge Depth	Welding Speed/Feed
1	900	0.10	55
2	900	0.12	60
3	900	0.15	65
4	900	0.20	70
5	1100	0.10	55
6	1100	0.12	60
7	1100	0.15	65
8	1100	0.20	70
9	1200	0.10	55
10	1200	0.12	60
11	1200	0.15	65
12	1200	0.20	70
13	1320	0.10	55
14	1320	0.12	60
15	1320	0.15	65
16	1320	0.20	70

TABLE 10.6
Tensile Testing Observation Result for AA7075 in UWFSW

S. No	Tool Speed (rpm)	Plunge Depth (mm/min)	Feed (mm/min)	Tensile Strength (MPa)
1	900	0.10	55	336
2	900	0.12	60	339
3	900	0.15	65	368
4	900	0.20	70	400
5	1100	0.10	55	410
6	1100	0.12	60	425
7	1100	0.15	65	396
8	1100	0.20	70	410
9	1200	0.10	55	399
10	1200	0.12	60	400
11	1200	0.15	65	418
12	1200	0.20	70	391
13	1750	0.10	55	325
14	1750	0.12	60	421
15	1750	0.15	65	399
16	1750	0.20	70	435

10.10.1 Determining the Single-to-Noise (S/N Ratio)

In this study, an L16 OA with 4 columns and 16 rows was used. This array can use four-level process parameters. Fourteen experiments using the L16 OA were performed to study. In this way, the effect of each selected factor on response and the S/N ratio for every factor were calculated. In this research work, hardness and tensile strength of the joints were identified as the responses for tensile strength "higher the better and normal the best".

In this research work, the effect of input parameters on tensile strength was analyzed. The S/N ratio for all the responses is analyzed for different variances of tensile strength. It clearly shows that the speed most significantly affects the tensile strength with a response means of 52.58 mm, a tensile strength of 480 MPa at 1750 rpm, and a plunge depth of 0.16 mm. It is clear from Table 10.7 that the optimal parameter has been established by the analyzing response curves of S/N ratio. It is found that the second level of speed is 1100 rpm, fourth level of plunge depth 0.16, and fourth level of feed 45 mm/min. Tensile strength test was performed to obtain the toughness and ductility of the welded joints in terms of elongation and strength, respectively.

This test was carried out using the tensile testing machine, which gives the tensile strength of the joint. This is expected in the SZ and TMAZ, which undergo plastic deformation and recrystallization, leading to grain refinement, as the strength of the welded joint increases without a decrease in the percent elongation. But, its formation of interaction in the joint greatly reduced the tensile strength. On the basis of rank of FSW parameters shown in Table 10.9, it was observed that the percent elongation is highly followed by rotational speed and plunge depth. Main effects plot for mean was also obtained, which is shown in Figure 10.8. It was observed that the optimum of the FSW parameters is found at 1750 rpm, with a welding speed of 45 mm/min and a shoulder diameter of 20 mm (1:4). High rotational speed causes an increase in the heat and size of coarse grain, which leads to a reduction in tensile strength.

Tool shoulder diameter plays an important role to achieve the tensile strength of the joint. Low diameter of tool dissipates less heat by decreasing heat-carrying capacity of tool. This restricts the proper distribution of the heat in the welding region and improves the tensile strength of the joint.

Larger value is the better criterion as it results in the better tensile strength (TS) in form of S2P4F4. ANOVA was applied to the signification of FSW parameters. Some diagnostic tests are required before performing ANOVA to validate the assumption made in ANOVA.

ANOVA was performed to assess the significance of FSW parameters and their relation with tool speed. The normal probability plot was obtained from Minitab 17 Software to check whether the residuals are normally distributed, which is shown in Table 10.9. It can be seen from Figure 10.10 that points lie either on the straight line or very close to it. Thus, the residuals are normally distributed and validated the ANOVA. The result of ANOVA shows that the rotational speed significantly affects the UTS followed by tool rotational speed and welding speed.

TABLE 10.7
ANOVA: UTS, Using Adjusted SS for Test, for UWFSW Process in Different Parameters

Tool Rotational Speed	Welding Speed	Plunge Depth	UTS (MPa)	SNRA 1
900	30	0.10	336	50.5268
900	35	0.12	339	50.6040
900	40	0.14	368	51.3170
900	45	0.16	400	52.0412
1110	30	0.10	410	52.2557
1110	35	0.12	425	52.5678
1110	40	0.14	396	51.9539
1110	45	0.16	410	52.2557
1320	30	0.10	399	52.0195
1320	35	0.12	400	52.0412
1320	40	0.14	418	52.4235
1320	45	0.16	425	52.5678
1750	30	0.10	325	51.7766
1750	35	0.12	421	51.8435
1750	40	0.14	399	52.7698
1750	45	0.16	480	52.9557

TABLE 10.8
Main Effects Plot for S/N Ratios of Rotational Speed versus Welding Speed

Level	Rotational Speed	Welding Speed	Plunged Depth
1	50.80	51.62	51.96
2	51.32	51.88	51.68
3	51.88	51.87	51.89
4	53.28	51.91	51.75
Delta	2.49	0.19	0.28
Rank	1	2	3

10.11   EFFECT OF ROTATIONAL TOOL SPEED ON UTS

It is evident from Figure 9.11 that the rotational speed is increased from levels 1 to 2, there is more increment in UTS due to USFSW process and again an increase in the rotational speed from levels 2 to 3, and there is a reduction in UTS value. But the value of UTS varies when the speed increases. The ultimate tensile strength, ductility, and net flow stress after yielding all increased with increasing welding speed in the UWFSW-welded joints.

Tool speed greatly affects the heat generation and material deformation, which leads to stirring and mixing of the BM for producing best quality of weld joints using

TABLE 10.9
ANOVA: Variance for UTS, Using Adjusted SS for Test

Source	DF	Seq SS	Adi SS	Adi MS	F	P	% Age Contribution
Rotational tool speed	3	4903.7	4903.7	1634.6	5.44	0.038	63.7
Welding Speed	3	572.7	572.7	190.9	0.63	0.619	7.44
Plunge depth	3	1017.2	1017.2	1017.2	339.1	1.13	12.21
Error	6	1204.4	1204.4	250.6	1.09	0.350	15.65
Total	15	7698					

Notes: The most significant parameters that effect on UTS, i.e., rotational tool speed, i.e., 63.7%, followed by plunge depth and welding speed.

$S=23.8537$, $R^2=69.78\%$, $R^2\ (adj)=24.45\%$.

TABLE 10.10
For Underwater FSW

Spindle Speed (rpm)	Plunge Depth (mm)	Welding Speed (mm/ min)	UTS1 (MPa)	UTS2 (MPa)	UTS3 (MPa)	UTS4 (MPa)	UTS5 (MPa)
900	0.10	30	341	339	345	353	361
900	0.12	35	340	347	353	362	368
900	0.14	40	357	359	340	346	378
900	0.20	45	349	351	345	345	357
1110	0.10	30	349	353	348	355	369
1110	0.12	35	390	387	367	330	400
1110	0.14	40	375	378	365	371	479
1110	0.20	45	360	365	372	375	383
1320	0.10	30	399	395	380	389	405
1320	0.12	35	385	379	400	417	478
1320	0.14	40	386	389	399	413	484
1320	0.20	45	400	410	401	472	460
1750	0.10	30	445	465	470	478	475
1750	0.12	35	465	456	400	484	488
1750	0.14	40	457	470	445	455	479
1750	0.20	45	480	479	489	479	480

cooling. Insufficient heat generated at lower tool speed (level 1) results in material movement leading to lower joint strength. On the other hand, higher rotational speed (level 3) generates more heat, which results in age-hardenable aluminum alloy and coarse grain growth, Therefore, rotational speed at level 2 generates enough heat to softening and proper consolidation of the BM with fine grain due to continuous recrystallization at level 3, 4.This results in defect-free joint with better UTS.

10.12   EFFECT OF WELDING SPEED ON UTS

The effect of welding speed with respect to UTS is depicted in Figure 10.9. It can be observed from Table 10.10 that ultimate tensile strength increases when the welding speed is increased from 1 to 2 levels. UTS decreases when welding speed is increase from 2 to 3 levels, and further, UTS is increased with increasing the welding speed from level 3 to level 4.

Welding speed mostly affects mechanical properties due to the heat generated in FSW process. Slow welding speed would cause more spread of heat over the weld length, which makes the welding process take a longer time. In this process, heat increases, which leads to high temperature of the weld zone. Higher temperature in the weld zone (WZ) results in an increase in the grain size, dissolution of precipitates, and an increase in coarsening of grain size, leading to an ultimate reduction in the tensile strength.

In UWFSW, high welding speed reduces heat generation due to cooling and increases flow stress, and it indicates the result in proper mixing of material during UWFSW process. In this research work, the welding speed at different levels might have sufficient heat for proper mixing of material in stir zone (SZ). The welding speed at 3 to 4 levels increases with sufficient temperature and time to obtain the best UTS, as shown in Figure 10.11.

10.13   EFFECT OF TOOL ROTATIONAL SPEED ON UTS

It is evident from Table 10.6 that as rotational speed increases from 1 to 2 levels, yield strength decreases, and as rotational speed increases, UTS increases from 2 to 3 and further increases the speed UTS from 3 to 4 levels. Increase in UTS in the last stage is relatively compared with the decrease in UTS in the former stage, which is shown in Figure 10.6.

Rotational tool speed affects the heat generated and material deformation, which results in stirring and proper mixing of the BM to produce the best quality of the welded joint. Heat generated at the starting point is not sufficient at lower rotational speed in level 1, leading to the low welded joint strength.

On the other hand, higher rotational tool speed from level 3 to level 4 generates more heat, which results in dissolution in the case of age-hardenable aluminum alloy and grain growth. Therefore, tool speed at level 3 generates sufficient heat for proper softening and consolidation of the BM with fine grains due to continuous recrystallization. This makes defect-free joint with better UTS.

10.14   IMPACT TEST

Impact test is carried out on a specimen using the Charpy impact test with a dimension of 55 mm×10 mm×6 mm and a deep notch of 2 mm at its center making an angle of 45° as shown in Table 10.11 and Table 10.12. The impact test machine is used to determine the properties of a metal such as toughness, hardness, and tensile strength during plastic deformation. It was observed that the joint showed a good toughness of 305 J for several experiments, but it was found better in UWFSW process; that is, the toughness was very low for few experiments (210 J).

The highest toughness from all the experiments was 305 J typically for other experiments having the toughness was 210 J. This is expected in the SZ and TMAZ, which undergo plastic deformation and dynamic recrystallization leading to grain refinement thereby increasing the strength of material. Increased strength and toughness of the metal after UWFSW process is due to the fine interlocking grain, if grain size found in coarse microstructure harmful to toughness. Better impact toughness as compared to the other weld parameters with 20 mm weld width at 1750 rpm is obtained.

Impact tests were carried out to determine toughness and ductility with respect to the strength of the material. Impact strength was measured by performing impact test. The obtained experimental reading of impact strength is shown in Table 10.11 and Table 10.12. The minimum toughness of AA7075 is 350 J at 900 rpm and 0.10 mm plunge depth, and maximum is 453 J at 1750 rpm and a plunge depth 0.16 mm.

TABLE 10.11

Observation Result of Impact Test for AA7075 Using Impact Testing Machine

S. No	Spindle Speed (rpm)	Toughness (J) at 0.10 Plunge Depth	Toughness (J) at 0.12 Plunge Depth	Toughness (J) at 0.14 Plunge Depth	Toughness (J) at 0.16 Plunge Depth	Toughness (J) at 0.18 Plunge Depth
1	900	210	215	235	245	240
2	1110	250	245	255	275	270
3	1320	265	260	270	295	280
4	1750	285	288	289	305	285
5	2220	230	220	245	265	260
6	2750	270	275	268	270	265

TABLE 10.12

Observation Result of Impact Test for AA7075 Using Impact Testing Machine

S. No	Feed (mm/min)	Toughness (J) at 0.10 Plunge Depth	Toughness (J) at 0.12 Plunge Depth	Toughness (J) at 0.14 Plunge Depth	Toughness (J) at 0.16 Plunge Depth	Toughness (J) at 0.18 Plunge Depth
1	30	220	218	235	245	255
2	35	260	265	285	290	292
3	40	270	285	290	275	265
4	50	288	280	288	285	270
5	60	235	230	265	250	260
6	65	270	280	282	290	285

It was observed that the joint found better toughness for many experiments; however, the toughness was very low (350 and 390 J) for few experiments (experiment nos. 1 and 6) due to the plunge depth. The highest toughness from experiment no. 5 was 453 J at a plunge depth of 0.16 mm at 1750 rpm due to proper bounding of the metal. It was nearly less than the toughness of the BM. This is quit in the SZ and "undergo severe plastic deformation and continuous recrystallization leading to the fine grains thereby increases the strength of the material, but, if re-precipitation of interlocking of the metal found during the FSW process, it greatly decreases the toughness.

From Table 10.12, it is very clear that while increasing the feed, the toughness is increased and while further increasing the feed, there is a decrement in the toughness. Table 10.12 shows the variation in feed due to the variation in factors. Figure 5.15 shows that the feed is increased from 1 to 2 levels. However, again while increasing the feed from 3 to 4 levels, the toughness of the joint is also increased. Maximum toughness at level 4 is 305 J at a plunge depth of .014. After 4 to 5 levels, the speed increases, and then the toughness of the joint decreases due to heat generation, which results in the increase in coarse grain size during the FSW process.

Increase in feed might have larger effects in toughness of the joint compared to the others. Thus, the effect of generated heat due to rotation tool takes over the effect of toughness of the joints, thereby reducing the strength of welded joints.

10.15 PREPARATION OF SAMPLES FOR MICROSTRUCTURE

Small pieces were cut from welding specimen samples using hacksaw and put into the circular ring, and then, the cold setting compound liquid paste was filled (cold setting compound mixed with water in the ratio of 1:2).

To get the solid piece, it was removed from the metal ring after 10 min.

Grinding was carried out using double-disc grinding machine at 300–350 rpm. AN interval time of 5 min for each sample with the different types of abrasive papers such as 800, 1000, 1200, 1500, and 2000 µm and 0.05 µm silica suspension was taken.

Finally, polishing was performed using an automatic double-disc polisher with velvet cloths as shown in Figure 10.9.

Samples were polished with diamond solution followed by alumina (alumina suspension) for microstructure examination. Light optical microscope (LOM) is used to examine the weld metal structure at 100× magnification.

Etching process was performed on polished samples to obtain the highlighting grain of weld joints with etching solution (2 ml HF+ 3 ml HCL+ 4 ml HNO3 + 190 ml H2O distilled water), and the microstructure was observed using the optical microscope.

10.16 MICROSTRUCTURE EVOLVED DURING UNDERWATER STIR FRICTION WELDING (UWSFW)

The microstructure zones of the UNFSW were pointed out in all welded joints by using an optical microscope. The weld joint made at 1320 rpm and a welding speed of 70 mm/min was analyzed for microstructure which has the highest hardness and tensile strength.

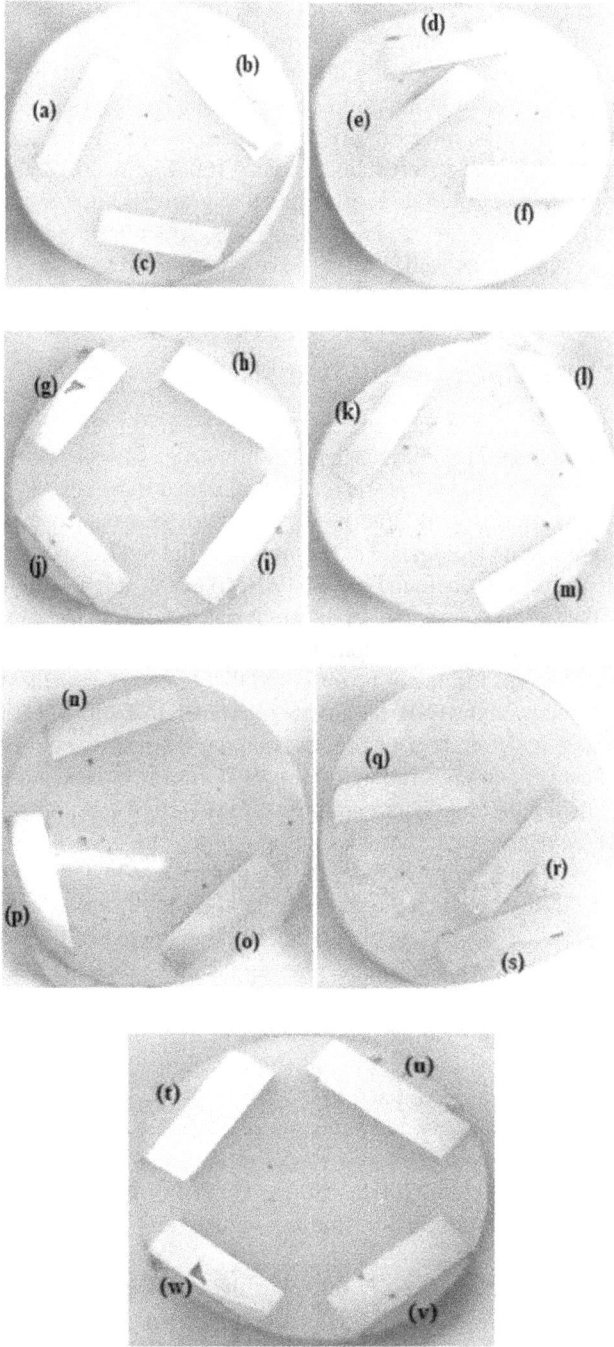

FIGURE 10.9 Samples for microstructure and Vickers hardness testing.

The grain size on NZ of AA7076 was better refined and continuous recrystallization due to the peak temperature. N. Martinez et al. investigated that the heat and
cooling rates influenced the nugget in different zones. The top surface on the nugget
has the better microstructure, fine grain size, and high strength.

Figure 10.12 shows the microstructure photographs of AA7075 aluminum alloy
weldments prepared under different conditions. From Figure 10.12, the microstructure of HAZ in AA7075 aluminum alloy has fine grain size due to a large number
of the inclusion of exhibit surface cavities or voids, which results in medium speed
and less heat generation as well, leading to low tensile strength of welded joint at
900 rpm

10.17 MICRO-HARDNESS DISTRIBUTION

Hardness is the resistance of the material to penetration. It is an important property
of the material compared to other mechanical properties such as strength. Hardness
measured for microconstituents of the material at a load less than 1 kg is referred to
as micro-hardness. In this work, micro-hardness testing was performed for measuring hardness of different biological phases present in the welded specimen. The average BM micro-hardness was found to be 149 HV Plots of micro-hardness data as a
position from the welded centerline for different process parameters are shown in the
figure 10.10 which shows the variations in micro-hardness with distance for different process parameters [13]. It can be observed that the micro-hardness distribution
considerably depends on the FSW parameters as shown in Table 10.13. The variation
of micro-hardness in different zones depends on the grain size and precipitation.

Optimum heat generation is required for proper mixing of base material and prevents the precipitate dissolution and grain growth [14–16]. Lower hardness values in
TMAZ and HAZ of the weld joint of aluminum alloys are due to grain growth and
precipitate dissolution. The hardness of the welded regions (TMAZ and WNZ) was
decreased compared to other regions. AA7075 alloy is the result of welding with
precipitate condition in which low plastic deformation in the BM by using HSS tool.
Welding with this condition leads to decrease in the hardness and tensile strength of
the welded joint.

Heat generated during welding process is showed by the neighboring regions of
SZ (i.e., HAZ and TMAZ) and leads to reduction in hardness. However, higher hardness values can also be observed at TMAZ due to the presence of high dislocation
density with fine grain caused by dynamic recrystallization.

Microstructure in different regions of the welded joints clearly shows that the
thermal cycle has considerably influenced the HAZ. In the THAZ, grain boundar-
ies grow because of the plastic deformation; hence, less heat is produced in this
region.

It is clear that the grain boundary separates the NZ from the TMAZ, leading to
the grain refinement in the NZ, and shows the improvement of mechanical properties
including hardness and strength in this region. Hardness in NZ and TMAZ/HAZ
relative to the BM, and the hardness decrease in HAZ/TMAZ due to coarsening of
the strengthening precipitate at higher temperatures, the high cooling rates preventing the amount of hardness in HAZ/TMAZ [19,20].

The dynamically recrystallized zone is the stirred zone, where the material has undergone plastic deformation resulting in fine grains. Micro-hardness of the FSW joints of aluminum alloys depends on the grain size, formation of precipitates, and dislocation density.

Micro-hardness distribution confirms that considerable grain refinements have occurred in the SZ, due to the dynamic recrystallization during FSW, which leads to higher micro-hardness in SZ than in the BM and HAZ, where the plastic deformation occurs. Hence, fine grains formed due to plastic deformation show high hardness values in the SZ [16,17]. Micro-hardness plots show difference in hardness distribution at AS and RS due to mechanical properties of the BM. Maximum hardness values were found in SZ, and minimum hardness was observed in TMAZ/HAZ. Hardness in NZ and TMAZ/HAZ decreases due to the dissolution of precipitates and the high cooling rates, preventing precipitation of phases and leading to a lower amount of hardness in HAZ/TMAZ.

TABLE 10.13

Micro-Hardness with Different Parameters Along with the Welding Bead for (UWFSW)

Distance	900 (rpm)	1000 (rpm)	1320 (rpm)	1750 (rpm)
0.5	131.9	137.9	138.9	136.8
1.0	130.3	136.3	136.3	137.0
1.5	135.2	131.2	137.1	131.3
2.0	133.0	136.0	138.0	134.0
2.5	137.7	138.7	148.7	142.7
3.0	138.8	142.7	146.7	141.9
3.5	136.1	139.2	140.0	146.8
4.0	135.6	136.9	137.0	137.4
4.5	138.0	138.3	138.4	135.0
5.0	137.3	136.5	136.5	138.4
5.5	139.5	135.4	137.0	134.5
6.0	137.8	138.4	136.0	138.8
6.5	134.0	133.7	135.6	135.3
7.0	135.0	136.2	131.0	134.0
7.5	137.3	136.7	130.0	136.6
8.0	130.8	134.3	136.1	135.0
8.5	133.9	135.7	138.3	131.7
9.0	132.6	133.2	135.6	133.6
9.5	136.6	132.3	133.5	131.2
10.0	137.0	135.2	131.4	135.7
10.5	130.0	129.2	129.3	127.8
11.0	129.4	129.7	127.4	129.7
11.5	131.8	137.6	134.8	136.9
12.0	137.3	136.1	137.6	135.4

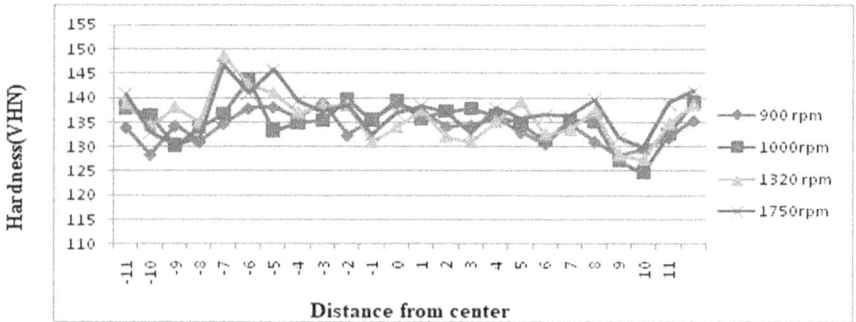

FIGURE 10.10   Micro-hardness distribution confirms at different welding speed and plunge depth.

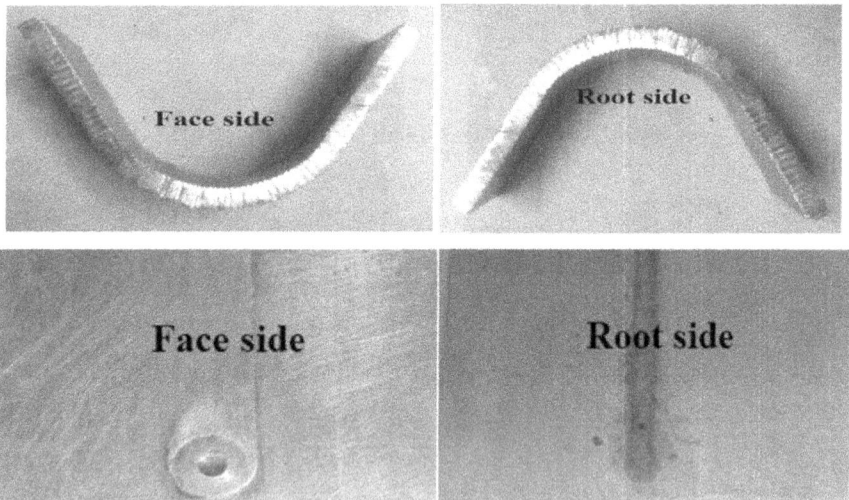

FIGURE 10.11   (a) Face side and (b) Root side. These bends are performed by the use of aqua fixture in FSW process.

10.18   SEM OF ALUMINUM ALLOY

Microstructure of the central portion of the weld joint of prepared specimens was analyzed to ensure the mixing of homogeneous distribution of particles during FSW process as shown in Figure 10.11. [17–19]. Microstructure characterization of the composites has showed proper uniform distribution and some amount of grain refinement in the specimens. Further, the hardness and tensile properties are higher in underwater FSW than in air medium. The microstructure and fracture were studied by optical microscopy (OM) and SEM in order to establish relationships between the quality of the fiber/aluminum interface bond and hence to link with mechanical properties of the composites [21,22]. The increase in Young's modulus is due to the hard basalt fiber and may be probably of homogenous distribution of basalt short fiber and the alignment of basalt short fiber parallel to the axis with minimum segregation in the alloys, which was confirmed from the microstructure studies.

10.19 FRACTOGRAPHY

Fractography of fractured sub-sized tensile specimen was performed to understand the fracture mechanism of the joints. Fracture of welded portion is described by the SEM micrographs at mid-section of the fractured tensile specimens as shown in Figure 10.12. Less populated deep and shallow dimples are observed in some joints typically shown in Figure 10.12. Void growth resulted in large deep dimples, implying involvement of the greater energy prior to fracture. Finely populated deep dimples are observed which implies typical ductile of fracture.

10.20 RESULTS AND DISCUSSION

Aluminum alloy AA7075-T6 is used in this research work and it is observed that the aluminum alloy AA7075-T6 has wide acceptance in industries such as aerospace and shipbuilding for lightweight component manufacturing as well as for structure building, because of its lightweight and high strength. It is analyzed that the grain size of microstructure is improved and the tensile strength is increased due to the flow of water on the surface of work piece during UWFSW because the heat input is controlled by using water. It is analyzed that the toughness of the specimen is decreased due to the underwater process. The microstructure of the three zones formed during UWFSW is analyzed, and as a result, it is observed that microstructure of all three zones during UWFSW is better than that of three zones during FSW.

During experimental investigation, it is analyzed that there are improvised mechanical properties due to the controlled heat in nugget. It is also analyzed that the strength of the welded joint in UWFSW is higher than that of welded joint in the FSW due to the low peak temperature.

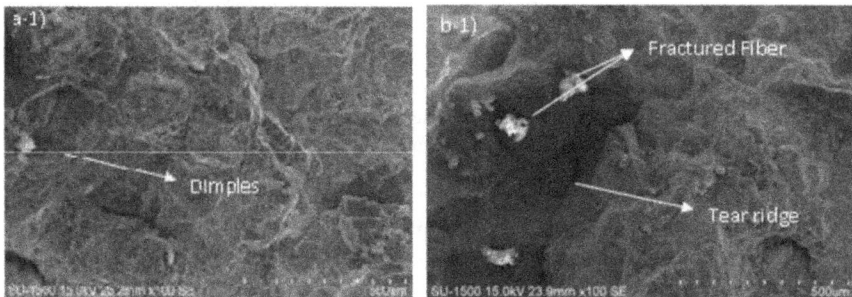

FIGURE 10.12   Fracture surfaces of samples tested for tensile test.

The microstructure analysis of UWFSW gives better tensile strength and hardness appearance as compared to the FSW process as shown in Figure 10.12. It is observed that the top surface on the nugget is having better microstructure, fine grain, and strength as per the microstructure analysis. In this chapter, the microstructure and mechanical properties are analyzed and discussed which show that both mechanical properties and microstructure are much better compared to simple FSW process. In this chapter, by applying Taguchi's optimization technique, it is observed that the better strength and fine microstructure are obtained at a tool speed of 1320 rpm, a plunge depth of 0.2 mm, and a welding speed of 70 mm/min.

10.21 CONCLUSION

Successful UWFSW of AA7075-T6 at a tool speed of 1750 rpm, a plunge depth of 0.2 mm, and a welding speed of 70 mm/min was found to be most appropriate weld joints.

The Taguchi design method is very efficient for optimizing welding parameters in performing the USFSW.

Tensile strength and hardness of AA7075-T6 are heavily influenced by the welding speed.

Maximum hardness is obtained at 148.7 HV and ultimate tensile strength 480 MPa.

Strength is improved in UWFSW-welded joint as compared to FSW due to low peak temperature.

REFERENCES

1. V.K. Dwivedi, A. Sharma, Influence on the tensile properties of AA7075-T6 under different conditions during friction stir welding process, *IOP Conference Series: Materials Science and Engineering* Vol. 691 (2019), p. 012001.
2. A. Sharma, V.K. Dwivedi, A comparative study of micro-structural and mechanical properties of aluminium alloy AA6062 on FSW and TIGW processes. *International Journal of Innovative Technology and Exploring Engineering* Vol. 8 (2019), pp. 337–342.
3. V.K. Dwivedi, A. Sharma, Experimental investigation of the mechanical properties and microstructure of AA 7075-T6 during underwater friction stir welding process. *International Journal of Engineering and Advanced Technology* Vol. 8 (2019), pp. 1289–1294.
4. J.M. Silcock, T.J. Heal, H.K. Hardy, Structural ageing characteristics of binary aluminum-copper alloys, *Journal of the Institute of Metals* Vol. 82, 2015, pp. 1953–1954.
5. F. Heirani, A. Abbasi, M. Ardestani, Effects of processing parameters on microstructure and mechanical behaviors of underwater friction stir welding of Al50alloy, *Journal of Manufacturing Processes* Vol. 25, 2017, pp. 77–84.
6. R. Burek1, D. Wydrzyński1, J. Andres1, A. Wrońska, Effect of tool eccentricity on microstructure and properties of FSW joints made of Al 7075-T6 Alloy, *Advances in Science and Technology Research Journal* Vol. 11, 2017, pp. 333–338.
7. S.A. Khodir, T. Shibayanagi, Microstruture and mechanical properties of friction stir welding of dissimilar AA-2024 and AA-7075 aluminum joints of AA2023-T3 and AA7075-T6, *Material Transactions* Vol. 48, 2007, pp. 1928–1937.

8. L.G. Eriksson, R. Larsson, Friction stir welding-now technology changing the rules of the game in Al construction, *Svetsaren* Vol. 15, 2001, pp. 323–451.

9. Sharma, A., Dwivedi, V.K. and Singh, Y.P., 2020. Effect on ultimate tensile strength on varying rotational speed, plunge depth and welding speed during friction stir welding process of aluminium alloy AA7075. Materials Today: Proceedings.

10. H.J. Zhang, H.J. Liu, L. Yu, Effect of water cooling on the performances of friction stir welding heat-affected zone, *Journal of Materials Engineering and Performance* Vol. 21–7, 2012, pp. 1182–1187.

11. S.S. Mirjavadi, M. Alipour, S. Emamian, S. Kord, A.M.S. Hamouda, P.G. Koppad, R. Keshavamurthy, Influence of TiO_2 nanoparticles incorporation to friction stir welded 5083 aluminum alloy on the microstructure, mechanical properties and wear resistance, *Journal of Alloys and Compounds* Vol. 712, 2017, pp. 795–803.

12. L.H. Wu, B.L. Xiao, D.R. Ni, Z.Y. Ma, X.H. Li, M.J. Fu, Y.S. Zeng, Achieving superior super plasticity from lamellar microstructure of a nugget in a friction-stir-welded Ti–6Al–4V joint, *Scripta Materialia* Vol. 98, 2015, pp. 44–47.

13. H. Liu, Y. Hu, C. Dou, D.P. Sekulic, An effect of the rotation speed on microstructure and mechanical properties of the friction stir welded 2060-T8 Al-Li alloy, *Materials Characterization* Vol. 123, 2017, pp. 9–19.

14. D. Li, X. Yang, L. Cui, F. He, H. Shen, Effect of welding parameters on microstructure and mechanical properties of AA6061-T6 butt welded joints by stationary shoulder friction stir welding, *Materials and Design* Vol. 64, 2014, pp. 251–260.

15. H.J. Aval, S. Serajzadeh, A.H. Kokabi, Evolution of microstructures mechanical properties in similar and dissimilar friction stir welding of AA5086 andAA6061, *Material Science and Engineering* Vol. 528, 2011, pp. 8071–8083.

16. H.I. Dawood, K.S. Mohammed, A. Rahmat, M.B. Uday, The influence of the surface roughness on the microstructures and mechanical properties of 6061 aluminum alloy using friction stir welding, *Surface & Coatings Technology* Vol. 270, 2015, pp. 272–283.

17. Q.D. Qina, B.W. Huanga, Y.J. Wub, X.D. Sua, Microstructure and mechanical properties of friction stir welds on unmodified and P-modified Al-Mg_2Si-Si alloys, *Journal of Materials Processing Technology* Vol. 250, 2017, pp. 320–329.

18. B.R. Sunil, G.P.K. Reddy, A.S.N. Mounika, P.N. Sree, Joining of AZ31 and AZ91 Mg alloys by friction stir welding, *Journal of Magnesium and Alloys* Vol. 3, 2015, pp. 330–334.

19. H. Shirazi, S. Kheirandish, M.A. Safarkhanian, Effect of process parameters on the macrostructure and defect formation in friction stir lap welding of AA5456 aluminum alloy, *Measurement* Vol. 76, 2015, pp. 62–69

20. F.F. Wang, W.Y. Li, J. Shen, S.Y. Hub, J.F. dos Santos, Effect of tool rotational speed on the microstructure and mechanical properties of bobbin tool friction stir welding of Al–Li alloy, *Material and Design* Vol. 86, 2015, pp. 933–940.

21. S. Singh, C. Prakash, P. Antil, R. Singh, G. Królczyk, C. I. Pruncu, Dimensionless Analysis for Investigating the Quality Characteristics of Aluminium Matrix Composites Prepared Through Fused Deposition Modelling Assisted Investment Casting. *Materials* Vol. 12, 2019, p. 1907.

22. C. Prakash, S. Singh, C. I. Pruncu, V. Mishra, G. Królczyk, D. Y. Pimenov, A. Pramanik, Surface modification of Ti-6Al-4V alloy by electrical discharge coating process using partially sintered Ti-Nb electrode. *Materials*, 12, 2019, p. 1006.

Index

For Product Safety Concerns and Information please contact our EU
representative GPSR@taylorandfrancis.com
Taylor & Francis Verlag GmbH, Kaufingerstraße 24, 80331 München, Germany